Lévy Statistics and Laser Cooli
How Rare Events Bring Atoms to R

Laser cooling of atoms provides an ideal case study for the application of Lévy statistics in a privileged situation where statistical models can be derived from first principles. This book establishes profitable connections between these two research fields, and demonstrates how the most efficient laser cooling techniques can be simply and quantitatively understood in terms of non-ergodic random processes dominated by a few rare events.

Lévy statistics is now recognized as the proper framework for analysing many different problems (in physics, biology, earth sciences, finance, etc.) for which standard Gaussian statistics is inadequate. Lévy statistics involves random variables with such broad distributions that the usual Central Limit Theorem no longer holds. Laser cooling allows atoms to be cooled to very low temperatures and brought to rest, and is a new research field with many applications. It provides a fruitful example of how approaches based on Lévy statistics can yield analytic predictions that can then be compared with both microscopic quantum optics treatments and experimental results.

The authors of this book are world leaders in the fields of laser cooling, light–atom interactions and statistical physics, and are also renowned for their clarity of exposition. Since the subject of this book embraces several different research areas, the authors have made every effort to ensure that it remains comprehensible to the non-specialist. They explain the important concepts of laser cooling and give an introduction to the concept of random walks and Lévy statistics, such that no detailed prior knowledge is required. This book will therefore be of great interest to researchers in the fields of atomic physics, quantum optics and statistical physics, as well as to engineers and mathematicians interested in stochastic processes. It will also be most useful for illustrating graduate courses on these topics.

FRANÇOIS BARDOU is a researcher at the Centre National de la Recherche Scientifique (CNRS) and currently works on problems in quantum stochastics at the Institut de Physique et de Chimie des Matériaux de Strasbourg. He received the 1995 Aimé Cotton prize (Atomic Physics prize of the French Physical Society) for his experimental and theoretical studies of laser cooling performed at the École Normale Supérieure de Paris. These studies, in collaboration with his co-authors, provided the basis for this book.

JEAN-PHILIPPE BOUCHAUD is a Senior Expert at the Service de Physique de l'État Condensé and at CEA-Saclay. In 1994 he founded his own company, 'Science and Finance', and continues to have diverse research interests which include statistical physics, granular matter and theoretical finance. In 1996 he won the

CNRS Silver Medal. He is in charge of a number of statistical physics and finance courses in various Grandes Écoles, Paris, and is the co-author of *Theory of Financial Risk* (Cambridge University Press, 2000).

ALAIN ASPECT is a Director of Research at CNRS and a Professor at the École Polytechnique, Palaiseau. After completing, in the early 1980s, a series of experiments on the foundations of quantum mechanics, he joined Claude Cohen-Tannoudji at the École Normale Supérieure to work on laser cooling of atoms. He is now head of the Atom Optics group of Institut d'Optique at Orsay and is the co-author of *Introduction to Lasers and Quantum Optics* (Cambridge University Press, in preparation).

CLAUDE COHEN-TANNOUDJI is Professor of Atomic and Molecular Physics at the Collège de France in Paris and was honoured with the Nobel Prize for Physics in 1997 for his work on the development of methods to cool and trap atoms with laser light. He is also the co-author of three other books: *Quantum Mechanics* (1992), *Photons and Atoms: Introduction to Quantum Electrodynamics* (1989), and *Atom–Photon Interactions: Basic Processes and Applications* (1998).

Lévy Statistics and Laser Cooling

How Rare Events Bring Atoms to Rest

FRANÇOIS BARDOU, JEAN-PHILIPPE BOUCHAUD,

ALAIN ASPECT and CLAUDE COHEN-TANNOUDJI

CAMBRIDGE
UNIVERSITY PRESS

CAMBRIDGE UNIVERSITY PRESS
Cambridge, New York, Melbourne, Madrid, Cape Town,
Singapore, São Paulo, Delhi, Tokyo, Mexico City

Cambridge University Press
The Edinburgh Building, Cambridge CB2 8RU, UK

Published in the United States of America by
Cambridge University Press, New York

www.cambridge.org
Information on this title: www.cambridge.org/9780521004220

First published 2002

A catalogue record for this publication is available from the British Library

Library of Congress Cataloguing in Publication Data

Lévy statistics and laser cooling: how rare events bring atoms to rest / François Bardou ... [*et al.*].
p.   cm.
Includes bibliographical references and index.
ISBN 0 521 80821 9 – ISBN 0 521 00422 5 (pb.)
1. Laser manipulation (Nuclear physics). 2. Laser cooling.
3. Atoms–Cooling. 4. Lévy processes. I. Bardou, François, 1967–
QC689.5.L35 L48 2001
539.7–dc21   2001025939

ISBN 978-0-521-80821-7 Hardback
ISBN 978-0-521-00422-0 Paperback

Contents

Foreword

Long ago, Paul Lévy invented a strange family of random walks – where each segment has a very broad probability distribution. These flights, when they are observed on a macroscopic scale, do not follow the standard Gaussian statistics. When I was a student, Lévy's idea appeared to me as (a) amusing, (b) simple – all the statistics can be handled via Fourier transforms – and (c) somewhat baroque: where would it apply?

As often happens with new mathematical ideas, the fruits came later. For example, É. Bouchaud proved that adsorbed polymer chains often behave like Lévy flights. In a very different sector, J.P. Bouchaud showed the role of Lévy distributions in risk evaluation. Now we meet a third major example, which is described in this book: cold atoms.

The starting point is a jewel of quantum physics: we think of an atom in a state of 0 translational momentum $p = 0$ (zero Doppler effect), inside a suitably prescribed laser field. For instance, with an angular momentum $J = 1$ we can have two ground states $|+\rangle$ and $|-\rangle$, and one excited state $|0\rangle$. The particular state $|+\rangle + |-\rangle$ has an admirable property: it is entirely decoupled from the radiation and can live for an indefinitely long time. It is thus possible to create a trap (around $p = 0$ in momentum space) in which the atoms will live for very long times: this so-called 'subrecoil laser cooling' has been a major advance of recent years. There are many statistical questions, concerning the resulting random flights in momentum space with alternate sequences of trapping and recycling. All the resulting effects in p space and in the time sequence can be measured and compared with statistical predictions inspired by the Lévy flights. (Here, the broad distributions are in the lifetimes, not in the size of the jumps.)

The present book summarizes these advances, incorporating a rare admixture of quantum physics and classical statistics. It is a meeting point for two cultures, each of them being represented by outstanding experts.

I am very impressed by this combination and by the clarity of the result. Both atomic physics and statistical physics integrate (roughly) a hundred years of culture. To extract what is needed from the two cultures and to make it accessible to a simple physicist was a real challenge. This joint group has done it. I am sure that many scientists will feel a special pleasure when reading the book – and that it will last a long time.

P.G. de Gennes
February 2001

Acknowledgements

We thank our colleagues from the statistical physics community as well as from the laser cooling community, in particular the members of the cold atoms group of the ENS, for very fruitful discussions and comments during the completion of this work. We have been greatly stimulated by the experimental results on subrecoil cooling obtained during the last 13 years. We are particularly indebted to Bruno Saubaméa and Jacob Reichel, whose PhD works on Lévy statistics applied to subrecoil cooling have made a significant contribution to the results presented in this book.

1

Introduction

This book deals with the important developments that have recently occurred in two different research fields, laser manipulation of atoms on the one hand, non-Gaussian statistics and anomalous diffusion processes on the other hand. It turns out that fruitful exchanges of ideas and concepts have taken place between these two apparently disconnected fields. This has led to cross-fertilization of each of them, providing new physical insights into the most efficient laser cooling mechanisms as well as simple and mathematically soluble examples of anomalous random walks.

We thought that it would be useful to present in this book a detailed report[1] of these developments. Our ambition is to try to improve the dialogue between different communities of scientists and, hopefully, to stimulate new, interesting developments. This book is therefore written as a case study accessible to the non-specialist.

Our aim is also to promote, within the atomic physics and quantum optics community, a way to approach and solve problems that is less based on exact solutions, but relies more on the identification of the physically relevant features, thus allowing one to construct simplified, idealized models and qualitative (and sometimes quantitative) solutions. This approach is of course common in statistical physics, where, often, details do not matter, and only robust global features determine the relevant physical properties. Laser cooling is an ideal case study, where the power of this methodology is clearly illustrated.

1.1 Laser cooling

During the last two decades, atomic physics has undergone spectacular progress based on a new experimental method, called laser cooling and trapping. By using

---

[1] Only a preliminary brief report of this work has been published [BBE94]. More detailed versions have been presented in an unpublished thesis work [Bar95] and in lecture notes [Coh96].

resonant or quasi-resonant exchanges of energy and momentum between atoms and laser light, it is now possible to obtain samples of atoms at temperatures in the microkelvin and even in the nanokelvin range, i.e. with velocities in the cm/s or in the mm/s range [Chu98, Coh98, Phi98]. Further cooling and increase of the density in phase space can then be achieved by using another recently developed method, called evaporative cooling. This has opened the way to a wealth of new investigations, ranging from ultrahigh resolution spectroscopy and atomic clocks to atomic interferometry and Bose–Einstein condensation (for a review of these fields see, for example, [AdR97, APS92, MeV99, BEC]).

In standard laser cooled atomic samples, called 'optical molasses', the ensemble of atoms interacting with appropriate sets of laser beams reaches a steady-state resulting from competition between two effects: damping of the atomic momenta due to a *friction* force originating from various types of velocity-dependent mechanisms ('Doppler' or 'Sisyphus' cooling) on the one hand and increase of the atomic momenta, or *momentum diffusion*, due to the fluctuations of the radiative forces, on the other. These fluctuations are associated with the random atomic recoils occurring in spontaneous emission processes which are generally unavoidable in any cooling scheme and which make the evolution of the atomic momentum look like a *random walk*. For a single spontaneous emission event, the recoil momentum of the atom has a magnitude (single photon recoil)

$$p_R = \hbar k, \tag{1.1}$$

where $\hbar k$ is the momentum of a photon (k is the optical wave-vector). It is therefore not surprising that, usually, the steady-state *rms* atomic momentum δp cannot be smaller than p_R: this is the so-called *single photon recoil limit* of laser cooling.

1.2 Subrecoil laser cooling

A completely different approach to laser cooling can be followed, which is not based on a friction force and where the single photon recoil no longer appears as a fundamental limit. The basic idea, presented in Fig. 1.1, is to create a 'trap' in momentum space, consisting of a small volume around $\mathbf{p} = \mathbf{0}$ ($\mathbf{p}$ denotes the atomic momentum), which the atoms can reach during their random walk and where they stay for a very long time, which increases indefinitely when $\mathbf{p} \to \mathbf{0}$. Such a situation is achieved by making the photon scattering rate (fluorescence rate) $R(\mathbf{p})$ vanish when $\mathbf{p} \to \mathbf{0}$. The random walk in momentum space slows down when $\mathbf{p}$ decreases and stops when $\mathbf{p} = \mathbf{0}$, so that atoms remain stuck in the neighbourhood of $\mathbf{p} = \mathbf{0}$. Up to now, this has been demonstrated by two methods, Velocity Selective Coherent Population Trapping (VSCPT) [AAK88] and Raman cooling [KaC92].

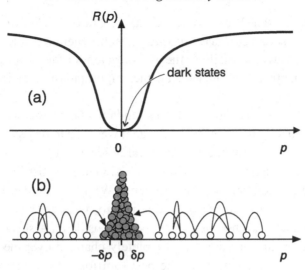

Fig. 1.1. Principle of subrecoil cooling. (a) The fluorescence rate $R(p)$ vanishes at momentum $p = 0$. (b) The atoms perform a random walk in p-space and accumulate in the vicinity of $p = 0$.

Cooling, i.e. an increase of the momentum space density in a narrow range around $\mathbf{p} = \mathbf{0}$, no longer results here from a friction force pushing the atoms towards $\mathbf{p} = \mathbf{0}$, but from a combination of two effects: momentum diffusion and vanishing of the jump rate of the random walk when $\mathbf{p} \to \mathbf{0}$. Another important difference between this and other cooling schemes is the absence of a steady-state value of the momentum distribution: the longer the interaction time θ, the narrower the range δp around $\mathbf{p} = \mathbf{0}$ in which the atoms can remain trapped during θ. Because of the absence of a steady-state and because of the existence of atomic characteristic times (trapping times) that can be longer than the observation time, here we will call such cooling *non-ergodic cooling*. It has also been called *subrecoil cooling* because nothing now prevents the atomic momentum spread δp reaching values smaller than the photon momentum $\hbar k$.

1.3 Subrecoil cooling and Lévy statistics

We present in this book a new general description of non-ergodic or subrecoil cooling in terms of a competition between trapping processes (i.e. the atom falls in the trap) and recycling processes (i.e. the atom leaves the trap and eventually returns to it). The fundamental feature which has stimulated the new approach presented in this book is that the distributions of trapping times and escape times can be very broad, so broad that the usual *Central Limit Theorem* (CLT) can no longer be applied to study the distributions of the total trapping time and of the total recycling time after N entries in the trap separated by N exits. We show in this book that the so-called *Lévy statistics*, which generalizes the CLT to broad distributions with power-law tails, is the appropriate tool for this problem and that it can provide quantitative results for all the important characteristics of cooled atoms.

Lévy statistics is an outcome of some fundamental mathematical work performed in the 1930's [Lev37, GnK54]. The goal was then to find stable distribution laws for the sum of N independent random variables, i.e. the distribution laws that keep the same mathematical form when $N \to \infty$. Gaussian distributions and Lévy distributions are the solutions of this problem. While the immense applicability of Gaussian distributions was recognized long ago, Lévy laws have been unduly ignored in the natural sciences and have been considered a sheer mathematical curiosity. However, the situation has completely changed over the last 15 years. Lévy statistics is now recognized as the best tool for studying many anomalous diffusion problems for which standard statistics are inadequate. Application fields include not only physics (anomalous diffusion, chaotic dynamics, mechanics of sandpiles, ...) but also finance, biology, etc. (see [BoG90, SZF95, Bak96, BCK97, Man97, Zas99, MaS99, PaB99, BoP00, CoR00, GoL01]). Lévy statistics can handle situations in which the standard deviation (or even the average value) of the studied random variable does not exist. It provides technical tools for performing calculations. Importantly, Lévy statistics implies properties that depart very strongly, not only quantitatively but even qualitatively, from usual statistical behaviour. For instance, when the average value of a random variable x is infinite, the sum $\sum_{i=1}^{N} x_i$ is no longer proportional to the number N of terms (usual law of large numbers), but a different *scaling* behaviour is obtained. This of course has dramatic phenomenological consequences, as we will see for the specific case of subrecoil laser cooling.

From the point of view of laser cooling, the study of subrecoil cooling by Lévy statistics turns out to be extremely fruitful. First, it allows one to extract the key ingredients of the cooling process from the relatively complicated microscopic description of the problem provided by atomic physics. Moreover, the statistical approach leads to unique analytical predictions for the asymptotic properties of the cooled atoms, independent of the details of the particular cooling scheme considered, as expected when one goes from a microscopic description to a statistical description.

From the point of view of statistics, this work can also be considered as a case study for the application of Lévy statistics in a privileged situation where the statistical model can be derived from first principles, developed analytically and, finally, precisely compared to microscopic theoretical approaches and to experimental results.

1.4 Content of the book

This book is intended for two different communities working in two different fields: atomic physics and quantum optics on the one hand and statistical processes on the other. We have thus considered it useful to include a summary of important results already known to each community, but not necessarily by both. These basic results are presented in Chapter 2 (laser cooling, see also Appendix A) and Chapter 4 (Lévy statistics), while Chapter 3 introduces the models that connect both fields. We then proceed in Chapters 5 and 6 with the derivation of the central results of this work. These results are then interpreted, discussed and extended in Chapters 7, 8 and 9. Appendices present several technical developments, either on Lévy statistics, or on subrecoil cooling processes.

More precisely, in *Chapter 2*, we recall some atomic physics results on laser cooling and subrecoil cooling. We point out the difficulties of an exact quantum treatment of subrecoil cooling using the generalized optical Bloch equations, and we present a more efficient quantum approach based on stochastic descriptions of the evolution of the wave function of the system, in terms of quantum jumps occurring at random times. This approach provides Monte Carlo simulations of the quantum evolution of the atomic momentum which allow one to describe, in a rigorous way, the cooling process by inhomogeneous random walks in momentum space with a momentum-dependent jump rate $R(p)$ vanishing for $p = 0$.

Such an approach suggests a simplified model where we make a partition between two classes of atoms: (i) the cold atoms, which are in a trapping volume in momentum space where the momentum $\mathbf{p}$ is close to zero, and which stay for a long time τ (trapping time) in this trapping volume (trapped atoms); (ii) atoms outside of the trapping volume, which make a random walk of duration $\hat{\tau}$ in momentum space, under the effect of radiation, until they come back again in the trapping volume ($\hat{\tau}$ is a first return time). We calculate in *Chapter 3* the probability distributions $P(\tau)$ and $\hat{P}(\hat{\tau})$ of the trapping times and first return times, and we show that in several important cases these distributions are broad distributions with power-law tails, for which Lévy statistics provides the relevant statistical treatment.

Chapter 4 summarizes the main results of Lévy statistics needed for the derivation and the interpretation of the results presented in this work. We will not give here all the detailed proofs, but rather emphasize the physical meaning of the results and the important differences between Lévy statistics and usual Gaussian statistics. More details may be found in [GnK54, Lev37, BoG90]. We also introduce a 'sprinkling distribution' which will be the basic tool for the calculations of the following chapters.

The concepts introduced in Chapter 4 are used in the following chapters for the derivation of quantitative predictions concerning laser cooling. We first study, in

Chapter 5, the proportion $f_{\text{trap}}(\theta)$ of cooled atoms after a cooling time θ, for the various cases considered in Chapter 3. We find not only the asymptotic behaviour of $f_{\text{trap}}(\theta)$, but the rate at which this asymptotic behaviour is reached when $\theta \longrightarrow \infty$. This allows us to give a first characterization of the efficiency of the cooling process. An important result of this chapter is also that $f_{\text{trap}}(\theta)$, defined here as an ensemble average, can be different from the corresponding time average. This clearly shows the *non-ergodic* character of the laser cooling process considered here.

A further step is achieved in *Chapter 6* by calculating the momentum distribution $\mathcal{P}(\mathbf{p})$ of the cold (trapped) atoms and the momentum distribution $\pi(p)$ along a given axis. We show that there is always in $\pi(p)$ a narrow peak whose width δp tends to zero when $\theta \longrightarrow \infty$, and the fraction of atoms contained in this peak is calculated in several important cases. The tails of $\pi(p)$ at large p are also studied. Their decrease is described by a power law, which shows that $\pi(p)$ is not a Gaussian distribution so that it is not possible, strictly speaking, to define a thermodynamic temperature. Finally, the increase of the density of atoms in momentum space and in phase space is evaluated for the various situations considered in this work.

The physical content of the results obtained in the preceding chapters is discussed in detail in *Chapter 7*. We re-interpret them in terms of rate equations describing a competition between a rate of entry in the trapping volume and a rate a departure. The rate of entry $S_R(t)$ is in fact nothing but the 'sprinkling distribution' introduced in Chapter 4. This enables one to interpret the behaviour of the height and of the peak of the momentum distribution. We also discuss in Chapter 7 a few other problems: the effect of a non-vanishing jump rate when $p \to 0$ and the connection between non-stationarity, non-ergodicity and broad distributions.

In *Chapter 8* we compare the analytical predictions of the statistical approach presented in this book with experimental results, as well as with predictions of microscopic theoretical approaches based on microscopic quantum treatments (stochastic wave function simulations or generalized optical Bloch equations). The excellent agreement between the various results gives us confidence in the approach developed in this work and in the approximations upon which it is based.

In *Chapter 9* we present an example of application of the approach developed in this book to a specific problem: the optimization of the height of the peak of cooled atoms. This brings into play both the insights and the technical results obtained in previous chapters and deepens our understanding of some properties of non-ergodic cooling.

We finally summarize in *Chapter 10* the main results derived in this book. We also mention a few possible extensions and a few open problems.

2
Subrecoil laser cooling and anomalous random walks

In this chapter, we first recall (in Section 2.1) a few properties of the most usual laser cooling schemes, which involve a friction force. In such standard situations, the motion of the atom in momentum space is a Brownian motion which reaches a steady-state, and the recoil momentum of an atom absorbing or emitting a single photon appears as a natural limit for laser cooling. We then describe in Section 2.2 some completely different laser cooling schemes, based on inhomogeneous random walks in momentum space. These schemes, which are investigated in the present study, allow the 'recoil limit' to be overcome. They are associated with non-ergodic statistical processes which never reach a steady-state. Section 2.3 is devoted to a brief survey of various quantum descriptions of subrecoil laser cooling, which become necessary when the 'recoil limit' is reached or overcome. The most fruitful one, in the context of this work, is the 'quantum jump description' which will allow us in Section 2.4 to replace the microscopic quantum description of subrecoil cooling by a statistical study of a related classical random walk in momentum space. It is this simpler approach that will be used in the subsequent chapters to derive some quantitative analytical predictions, in cases where the quantum microscopic approach is unable to yield precise results, in particular in the limit of very long interaction times, and/or for a momentum space of dimension D higher than 1. This approach also has the advantage of showing, from the beginning, that anomalous random walks and Lévy statistics are deeply involved in subrecoil laser cooling.

2.1 Standard laser cooling: friction forces and the recoil limit

2.1.1 Friction forces and cooling

Laser cooling consists of using resonant exchanges of linear momentum between atoms and photons for reducing the momentum spread δp of an ensemble of atoms. Achieving the smallest possible value of δp has led to spectacular developments in various research fields such as metrology, high resolution spectroscopy, atom optics, atomic interferometry, Bose–Einstein condensation of atomic gases, ...

In the most usual laser cooling schemes, an ensemble of atoms interacts with laser beams with suitable polarizations, intensities and frequencies, so that the atomic momenta are damped. In the so-called 'Doppler cooling' scheme for instance [HaS75, WiD75], this damping is due to an imbalance between opposite radiation pressure forces, induced by the Doppler effect and resulting in a net force opposed to the atomic motion. In the more efficient 'Sisyphus cooling' scheme [DaC89, UWR89, CoP90], atoms run up potential hills (where they decelerate) more frequently than they run down, with the net result of a decrease of atomic kinetic energy.

In all these situations, the cooling effect can be expressed by a friction force which, around $\mathbf{p} = \mathbf{0}$, is proportional to the atomic momentum $\mathbf{p}$ with a negative coefficient [Coh90]. When the friction coefficient is large, atoms seem to move in a very viscous medium, and this kind of situation is called 'optical molasses' [CHB85, LPR89].

Momentum damping by a friction force is a dissipative process, necessarily associated with some fluctuations. In the laser cooling schemes considered above, the fluctuations are due to spontaneous emission of fluorescence photons which can be emitted at random times and in random directions, resulting in a random fluctuating component of the momentum exchanged between the atom and the radiation field. Laser cooling mechanisms associated with a friction force therefore give rise to a random walk of the atomic momentum $\mathbf{p}$. As in usual Brownian motion, such a random walk can be characterized by a drift of the atomic momentum towards $\mathbf{p} = \mathbf{0}$ due to the friction force (damping of the mean momentum), and by a momentum diffusion, due to the randomness introduced here by spontaneous emission. Competition between friction and diffusion eventually leads to a steady-state, where the momentum distribution can be characterized by a stationary probability distribution of half-width δp. However, even in the steady-state, fluorescence cycles never cease, and the random walk in momentum space never stops.

Even in the steady-state of the cooling process, laser cooled gases are generally not in thermal equilibrium and, as such, cannot be characterized by a well defined thermodynamic temperature. It is, however, convenient to express the half-width δp of the distribution using an 'effective temperature' T defined by

$$\frac{1}{2} k_{\mathrm{B}} T = \frac{\delta p^2}{2M},\tag{2.1}$$

where k_{B} is the Boltzmann constant, M is the atomic mass and δp is the half-width at $\mathrm{e}^{-1/2}$ of the maximum of the one-dimensional momentum distribution. When the momentum distribution is Maxwell–Boltzmann, the definition of eq. (2.1) coincides with the usual definition of statistical mechanics. Throughout this book, we will use notations T and δp as defined by eq. (2.1).

2.1.2 The recoil limit

What is the minimum temperature that can be reached by standard laser cooling mechanisms? Since fluorescence cycles never cease and involve spontaneously emitted photons which communicate to the atom a random recoil, it seems difficult to control the momentum spread δp to better than the 'recoil limit' $\hbar k$, the momentum of a single photon. In terms of temperature, the recoil limit reads:

$$T \geq T_R, \qquad (2.2)$$

where the 'recoil temperature' T_R is defined by

$$T_R = \frac{\hbar^2 k^2}{M k_B}. \qquad (2.3)$$

The quantity $k_B T_R/2$ is the kinetic energy given to an atom at rest by the absorption or the emission of a single photon. For most atoms, the recoil temperature is of the order of 1 microkelvin (μK). Optimized Sisyphus cooling leads to temperatures close to the recoil limit [CaM95].

2.2 Laser cooling based on inhomogeneous random walks in momentum space

2.2.1 Physical mechanism

It is also possible to achieve laser cooling, i.e. to accumulate atoms around the origin $\mathbf{p} = \mathbf{0}$ of the momentum space, without any friction force. Rather than pushing the atoms towards $\mathbf{p} = \mathbf{0}$ (drift of the momentum random walk associated with the friction), one resorts to an inhomogeneous diffusion coefficient, vanishing around $\mathbf{p} = \mathbf{0}$, so that the random walk slows down in this region, where the atoms pile up. Although random walks seem at first sight to be less efficient at reducing momenta than the deterministic trend provided by friction, it is these random walks that will enable one to circumvent the recoil limit.

Such a situation is achieved when the fluorescence rate R, at which photons are spontaneously re-emitted, depends on $\mathbf{p}$ and exactly vanishes for atoms with zero momentum $\mathbf{p} = \mathbf{0}$ (Fig. 1.1a).

The consequence of the vanishing of the fluorescence rate $R(\mathbf{p})$ around $\mathbf{p} = \mathbf{0}$ is that ultracold atoms ($p \simeq 0$) no longer undergo the random recoils which would be due to spontaneous emissions. They are in some sense protected from the 'bad' effects of light. On the other hand, atoms with $p \neq 0$ can absorb light. The random recoil due to the re-emitted photons modifies in a random way their momentum, which can move closer to zero, or farther from zero, after each fluorescence cycle. The **p**-dependent fluorescence rate of Fig. 1.1a is thus at the origin of an inhomogeneous random walk in **p**-space, with a **p**-dependent jump rate $R(\mathbf{p})$, which eventually transfers atoms from the $p \neq 0$ absorbing states into the $p \simeq 0$ non-absorbing 'dark' states where they remain trapped and pile up (Fig. 1.1b).

A simple analogy can help understanding such a physical mechanism. Consider sand grains in a Kundt tube where a resonant acoustic standing wave is excited. The sand grains are vibrating and moving along the axis of the standing wave in an erratic way, except at the nodes of the standing wave, where there is no sound vibration to excite them. After a certain time, the sand grains accumulate at the nodes of the standing wave. In both cases, cooling without friction and Kundt's tube, we have an inhomogenous random walk, i.e. a random walk with a jump rate varying with the location of the particle and vanishing at certain places. For cooling, however, the random walk is in momentum space and the jump rate is momentum-dependent, whereas in a Kundt's tube, the sand grains move in real space, and the jump rate is position-dependent.

The fact that the cooling processes considered here do not rely on a friction force does not mean of course that they could not benefit from the presence of such a friction force. Without friction, the cooling relies on the efficiency with which a pure random walk can bring an atom near $\mathbf{p} = \mathbf{0}$. Such an efficiency decreases dramatically with the number D of dimensions. In one dimension, every particle returns often to the origin of momentum space, although it may take a very long time; in two dimensions, this return takes a much longer time and in three dimensions returning to the origin becomes even more unlikely. It is thus very useful to supplement the momentum random walk with a friction force, producing a drift that tends to push the atoms towards the origin of momentum space [MaA91]. This clearly improves the accumulation process. All subrecoil cooling schemes that have been implemented offer the possibility of an efficient friction force.

2.2.2 How to create an inhomogeneous random walk

Up to now, two methods of laser cooling based on the inhomogeneous random walk presented above have been proposed and demonstrated: Velocity Selective Coherent Population Trapping (VSCPT) [AAK88] and Raman cooling [KaC92]. In this book, we do not present a detailed description of these two methods[1], which can be found elsewhere [AAK89, Coh90, Ari91, OlM90, MaA91, KaC92, Rei96,

---

[1] However, Appendix A gives a detailed derivation of the parameters of inhomogeneous random walks corresponding to VSCPT and Raman cooling.

RSC01]; we just indicate how these methods fit into the scheme discussed in this work.

In Coherent Population Trapping, an atom with several ground state sublevels interacts with a set of quasi-resonant laser beams, such that there exists a particular linear superposition of these ground state sublevels where the atom does not interact with the laser light. This cancellation of the coupling is due to a destructive quantum interference between absorption amplitudes, and it leads to the accumulation of the atoms into the uncoupled state where the atoms cease to fluoresce ('dark state') and become trapped [AGM76, ArO76]. A careful analysis taking into account the quantization of atomic motion [AAK89, Coh90, Coh96] shows that, for most laser configurations, the cancellation of the fluorescence rate actually depends on a generalized atomic momentum $\mathbf{p}$ (which is now a quantum number) and that situations exist where $R(\mathbf{p})$ exactly cancels at $\mathbf{p} = \mathbf{0}$. The fluorescence rate thus becomes 'Velocity Selective'. In this VSCPT process, one can show that $R(\mathbf{p})$ varies quadratically with $\mathbf{p}$ around $\mathbf{p} = \mathbf{0}$. On the other hand, the behaviour of $R(\mathbf{p})$ at large $\mathbf{p}$ depends on the laser configuration, but it usually saturates and decreases asymptotically as p^{-2} because of the Doppler shift. Most VSCPT schemes give rise to a friction force for a proper sign of the detuning between the laser frequency and the atomic frequency [SHP93, MDT94, WEO94, HLO00]. Notice, however, that the initial VSCPT scheme [AAK88] involves no such force and appears to be an example of cooling due purely to an inhomogeneous random walk.

In Raman cooling, an atom with two hyperfine ground state sublevels interacts with a sequence of laser pulses leading to population transfers between the ground state sublevels, through stimulated and spontaneous Raman processes. Here also, this process can be made momentum-dependent, and it is possible to choose sequences of pulses such that atoms stop interacting with the lasers at $\mathbf{p} = \mathbf{0}$. The $\mathbf{p}$-dependence of the fluorescence rate $R(\mathbf{p})$ around $\mathbf{p} = \mathbf{0}$, as well as for large values of $\mathbf{p}$, can even be tailored almost at will [KaC92, RBB95]. Moreover, a friction force is readily implemented by an appropriate choice of the directions of the laser beams used for the stimulated Raman transitions [KaC92].

2.2.3 Expected cooling properties

The exact vanishing of the random walk jump rate $R(\mathbf{p})$ at $\mathbf{p} = \mathbf{0}$, which is the very basis of the non-standard cooling mechanisms considered here, has important consequences.

A first consequence of the vanishing of $R(\mathbf{p})$ around $\mathbf{p} = \mathbf{0}$ is the absence of a steady-state, even at arbitrarily long interaction times θ. Indeed, let us consider the characteristic evolution time $\tau(\mathbf{p})$ defined by

$$\tau(\mathbf{p}) = R(\mathbf{p})^{-1}, \tag{2.4}$$

i.e. the mean time for an atom with momentum $\mathbf{p}$ to undergo a fluorescence cycle, and let us compare it to the 'interaction time' θ, a key quantity in this study, which represents the duration of the interaction between the atoms and the laser beams that cool them. Since $R(p) \underset{p\to 0}{\to} 0$, one has $\tau(p) \underset{p\to 0}{\to} \infty$. Therefore, however long the interaction time θ may be, there exists atomic evolution times $\tau(\mathbf{p})$ longer than θ, namely the ones corresponding to $p < p_\theta$ where p_θ is defined by

$$R(p_\theta)\,\theta = 1. \tag{2.5}$$

We will show in this work that these unbounded evolution times can introduce a fundamental non-ergodicity in the problem, hence the denomination 'non-ergodic cooling' used in this book.

As a second consequence, the single photon recoil $\hbar k$ is no longer a fundamental limit for cooling. In this study, we will show that the radius δp of the volume around $\mathbf{p} = \mathbf{0}$ where the atoms are accumulated depends only on the shape of $R(\mathbf{p})$ around $\mathbf{p} = \mathbf{0}$, and on the interaction time θ between the atoms and the lasers. The radius δp will be shown to be of the order of p_θ, defined according to eq. (2.5) as the smallest momentum for which the probability of a jump during θ is not negligible. Since $R(p)$ increases monotonically with p, this relation leads to values of $\delta p \simeq p_\theta$ which decrease indefinitely with θ, and thus become smaller than the recoil $\hbar k$ for long enough interaction times. This is how the recoil limit has been overcome in laser cooling experiments based on the schemes described here [AAK88, KaC92, LBS94, LKS95, RBB95].

The key role played by the very long sojourn times around $\mathbf{p} = \mathbf{0}$ is the basic ingredient of cooling by inhomogeneous random walks. In fact, the probability distribution of the sojourn times τ between successive jumps, has such long tails for large values of τ, that the variance or even the average of τ may not exist. This is why the inhomogeneous random walks associated with these cooling mechanisms are anomalous, and cannot be treated by standard statistics.

2.3 Quantum description of subrecoil laser cooling

2.3.1 Wave nature of atomic motion

The description of subrecoil cooling given in Section 2.2 is oversimplified. In the subrecoil regime, the external degrees of freedom, i.e. the position x and the momentum p of the atomic centre of mass, must be treated quantum-mechanically (we take one-dimensional notations for simplicity). This is due to the fact that the condition $\delta p < \hbar k$ is equivalent to $h/\delta p > h/(\hbar k)$, i.e. to $\xi > \lambda_L$ where $\xi = h/\delta p$ is the atomic spatial coherence length and $\lambda_L = h/(\hbar k)$ is the laser wavelength. In other words, below the recoil limit, the spatial coherence length of atomic wave packets becomes larger than the wavelength of the lasers which are used to cool the atoms, so that it becomes impossible to make the semiclassical approximation consisting of a classical treatment of the position of the atomic centre of mass [Coh90]. In this regime, the wave nature of the atomic motion can no longer be ignored, and the description of the atomic state must include not only internal quantum numbers g_m, but also external quantum numbers q labelling the motion of the centre of mass. Moreover, we want to describe the evolution of an ensemble of atoms submitted to dissipation.

In order to take into account both the wave nature of the atomic motion and the dissipation, the quantum description generally uses a density matrix σ, the elements of which $\langle g_m, q | \sigma | g_{m'}, q' \rangle$ are labelled by internal and external quantum numbers. The equations of motion of this infinite set of density matrix elements (including those which involve the Zeeman sublevels of the excited state manifold e) are the so-called Generalized Optical Bloch Equations (GOBE).

2.3.2 Difficulties of the standard quantum treatment

Because of their complexity, it is in general impossible to obtain analytical solutions of the GOBE. Furthermore, in the subrecoil laser cooling mechanisms considered here, there is no steady-state solution, and it is thus difficult to make simple predictions concerning the long time regime where these mechanisms are most interesting. There are, however, a few exceptions for simple one-dimensional laser configurations, where analytical predictions for asymptotic behaviour have been obtained [AlK92, AlK93, AlK96, SSY97]. Unfortunately, it seems to be extremely difficult to extend these methods to more general laser configurations and to higher dimensions.

A numerical integration of the GOBE also raises serious practical problems in the limit of long times. Suppose that one chooses momentum eigenvalues p for labelling the state of the centre of mass. To make numerical calculations, one must discretize p over a grid which has to be finer and finer as time grows, because of the appearance of a narrow peak in the momentum distribution, with the width p_θ tending to zero when θ tends to infinity. But in the same limit (and in the absence of friction), the momentum diffusion also spreads the possible values of p over a larger and larger interval which grows with increasing θ. It follows that the number of equations to be numerically solved becomes prohibitively large, especially in dimensions larger than one. It is therefore impossible to make precise calculations for long times, and to obtain information concerning, for example, the proportion of cooled atoms contained in the narrow peak of the momentum distribution (see, for example, Section 6.F of [AAK89]).

2.3.3 *Quantum jump description. The delay function*

A deeper understanding of VSCPT can be obtained [Coh90, CBA91, Coh92] by applying to this problem the method of the delay function which has been developed for analysing photon correlations in resonance fluorescence [Rey83, CoD86], or the quantum jumps observable in the fluorescence light emitted by a single trapped ion [CoK85]. The initial motivation for this new investigation of subrecoil cooling was to overcome some difficulties of the numerical treatment of VSCPT by the GOBE at long interaction times θ. But the most important reason for briefly describing the results obtained by this approach is that they will lead us to the statistical model studied in this book, establishing the connection with Lévy statistics.

In most quantum optics problems, where an atom is driven by coherent laser fields and where the main source of dissipation comes from spontaneous emission, it is possible to describe the evolution of a single atom as a stochastic evolution consisting of a sequence of coherent evolution periods, involving absorption and stimulated emissions, separated by quantum jumps consisting of spontaneous emission processes which occur at random times. Such an evolution can also be interpreted in terms of a radiative cascade of the total system {atom + laser photons}, the so-called 'dressed atom' (see, for example, Chapter VI of [CDG88]). In this picture, each quantum jump is associated with a spontaneous transition of the dressed atom between two adjacent manifolds, by spontaneous emission of a fluorescence photon, and it puts the atom in a well defined state of the new manifold, correlated to the state of the spontaneously emitted photon. Between two quantum jumps, the dressed atom evolves within a given manifold. This coherent evolution is due to absorption and stimulated emission of laser photons by the atom, and it is described by a non-Hermitian effective Hamiltonian H_{eff}, which includes the effect of the damping due to spontaneous emission.

In this picture, it is natural to introduce the probability distribution $W(\tau)$ of the time intervals τ between two successive spontaneous emissions of photons by the same atom. Such a function, which is called the 'delay function' or the 'waiting time distribution', can be calculated by diagonalizing H_{eff} (see Appendix A.1). From $W(\tau)$ one can then deduce all the statistical properties of the random sequence of fluorescence photons. In particular, one can use $W(\tau)$ to make a Monte Carlo simulation of the random sequence of quantum jumps [CBA91, ZMW87]. Between two successive quantum jumps, the atom is described by its coherent evolution in a given manifold. Such a sequence of coherent evolutions separated by random jumps constitutes a quantum jump simulation, and averaging over an ensemble of such simulations yields a quantum jump description of the process. Note that although the Monte Carlo drawings are based on usual (definite positive) probability laws such as $W(\tau)$, the quantum jump description is not classical, and contains all the quantum features of the evolution since these probability laws have been obtained from a quantum treatment of the problem. The quantum jump description thus provides a statistical description of all the interesting quantities.

We have performed quantum jump simulations of VSCPT, choosing an atomic transition and a one-dimensional laser configuration (corresponding to a real experiment [AAK88]), simple enough for allowing the delay function $W(\tau)$ to be calculated by diagonalizing a 3×3 matrix (only three internal states are involved) [CBA91, BBE94, Bar95]. In such a simple case, which has also been studied by using optical Bloch equations [AAK89], the coherent evolution periods take place within a three-dimensional atomic subspace $\mathcal{F}(p)$ (family of three atomic quantum states which are coupled by absorption and stimulated emission) labelled by using an index p which has the meaning of a momentum (the momentum of the system {atom + laser photons}). We call $W_p(\tau)$ the delay function obtained by diagonalizing H_{eff} within the subspace $\mathcal{F}(p)$. In a spontaneous emission process, the system changes randomly from a family $\mathcal{F}(p)$ to another one $\mathcal{F}(p')$ with $|p - p'| \leq \hbar k$, and we have to calculate a new delay function $W_{p'}(\tau)$. The time of the quantum jump which makes the atom leave $\mathcal{F}(p)$, and the corresponding change of atomic state, are obtained by a Monte Carlo procedure based on the delay function, and on the probability law of the change of atomic state, derived from a quantum analysis of the situation. Finally, the result of each simulation can be simply presented by giving the sequence of the constant values of p between two successive quantum jumps, and the times of the quantum jumps (see Fig. 2.1).

2.3.4 Simulation of the atomic momentum stochastic evolution

We have represented in Fig. 2.1, which is taken from [BBE94], an example of the stochastic evolution of the momentum p of an atom undergoing VSCPT, given by the quantum jump simulation described above.

At certain random times, the atom emits a photon and p changes abruptly. Between two successive spontaneous emissions, p remains constant. This simulation allows us to make the connection between the quantum description of VSCPT and the key ingredients of the cooling processes studied in this work. It clearly appears in Fig. 2.1 that the smaller p, the longer the delay τ between two successive spontaneous emissions: this is the principle of the new cooling schemes involving inhomogeneous random walks.

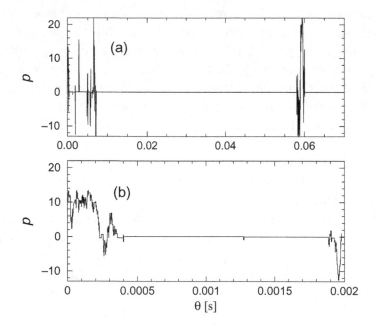

Fig. 2.1. (a) Example of a momentum random walk resulting from a quantum jump simulation of subrecoil cooling of metastable helium atoms. The unit of atomic momentum p is the momentum $\hbar k$ of the photons. The zoom (b) of the beginning of the time evolution is statistically analogous to the evolution at large scale. A striking point, typical of Lévy flights, is the fact that, for all time scales, the few longest time intervals dominate the evolution. Parameters: $\Omega_1 = 0.3\Gamma$ and $\delta = 0$ (see Appendix A for details).

There is another striking feature in Fig. 2.1, which was the starting point of the presented approach: the random sequence of time intervals is clearly dominated by a single term, the longest one, which is of the order of the total observation time. This feature gives an anomalous character to the random walk along the time axis. As we will see below, this statistically anomalous behaviour is the heart of the efficiency of subrecoil cooling and requires special statistical methods for its description.

2.3.5 *Generalization. Stochastic wave functions and random walks in Hilbert space*

More general one-dimensional configurations, where the delay function cannot be easily calculated, have also been investigated [MDT94] by using the so-called 'Monte Carlo Wave Function' (MCWF) method [DCM92, MCD93, MoC96], which in fact consists of a numerical calculation of the delay function, step by step. One can still consider families $\mathcal{F}(p)$ within which coherent evolution takes place between two successive quantum jumps. For instance, in general laser configurations used in one-dimensional VSCPT, the laser electric field, which results from the superposition of two counterpropagating laser waves, varies periodically in space. The corresponding periodic optical potential in which the atom moves gives rise to a band structure of Bloch states [CaD91]. One can show [Coh96] that a given family $\mathcal{F}(p)$ is nothing but the set of Bloch states having the same quasi-momentum p. This gives a physical meaning to the index p labelling the families $\mathcal{F}(p)$. But these families are now atomic subspaces of infinite dimensions, and it is no longer possible to calculate the delay function by a simple diagonalization of a low dimension matrix. However, the resulting evolution can still be simply represented by a sequence of quantum jumps separated by coherent evolutions in well defined families labelled by a constant number p. Note again that there is no classical approximation in the above procedure, which retains all the quantum features of the problem.

The delay function approach and the MCWF method are both stochastic approaches belonging to a more general theoretical framework which is now being developed for analysing dissipative quantum optics problems and which is based on the idea of *stochastic wave functions*. There are a number of slightly different schemes [Car93, DZR92] which all share two basic ideas that distinguish stochastic wave functions from the traditional approach of optical Bloch equations. First, the atoms are no longer described by a density matrix but rather by an ensemble of wave functions. Second, the time evolution of these wave functions is not a fully continuous deterministic process as is the case for Bloch equations, but rather a sequence of coherent evolutions (continuous and deterministic) interrupted at random times by random instantaneous quantum jumps corresponding to spontaneous emissions.

Such a general stochastic wave function approach, which can be shown to be mathematically equivalent to Bloch equations, presents several interesting new features.

First, it provides a possible description of the quantum dissipative evolution of individual particles. Thus, by following the evolution of a single wave function, one gets a physical intuition of the quantum behaviour of a single atom, that would be harder to derive from the density matrix used in GOBE[2].

Second, the stochastic wave function approach can be efficiently implemented numerically. Indeed, in a problem where a continuous variable (like the momentum) is discretized in N steps with $N \gg 1$, it is much easier to perform calculations on wave functions of size N than on the corresponding density matrix of size

---

[2] For instance, this has been very useful for understanding the so-called 'dark periods' appearing in the fluorescence emitted by a single trapped ion [CoD86], or important features of certain schemes of amplification without inversion [CZA93].

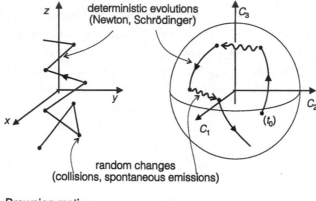

(a)   Brownian motion
(real space (x,y,z))

(b)   Stochastic wave functions
(Hilbert space (C_1, C_2, C_3))

Fig. 2.2. Random walks: (a) of a classical particle in position space, under the influence of collisions with gas molecules interrupting free flights; (b) of a stochastic wave function in a Hilbert space (here three-dimensional), under the influence of spontaneous emissions interrupting Hamiltonian evolutions. Note that the wave function is normalized and therefore its motion is restricted to a sphere.

$N \times N$. This has proved to be crucial for the study of three-dimensional laser cooling [CaM95]. Another example is the analysis of the original one-dimensional VSCPT scheme: the explicit use of the delay function enables one to carry out calculations up to time scales 10 000 times larger than previous works with GOBE. Furthermore, the Monte Carlo algorithm exhibits multiscale properties that are especially well adapted for non-ergodic processes [Bar95]. Indeed the delay function allows one to choose the next spontaneously emitted photon in a single calculation step, even if the corresponding delay is very long. Since non-ergodic cooling is characterized by extremely long delays, this is a crucial advantage. In the case of more complex schemes, in which a step-by-step time integration replaces the delay function, the stochastic wave functions are still more tractable than the GOBE.

Third, stochastic wave functions provide a new description of quantum dissipative processes, which is very similar to a classical random walk (see Fig. 2.2). Indeed, between two spontaneous emissions, the atomic wave functions obey a coherent deterministic evolution, just as a Brownian particle obeys a free flight motion between two collisions. Then, the spontaneous emission is considered as instantaneous and random, occurring at random times and in random directions, just as collisions experienced by a Brownian particle. We are thus led to the important idea that a stochastic wave function performs a random walk in Hilbert space [BrP96]. Of course, the random walk of the wave function is more complicated than the random walk of a classical Brownian particle which is performed in a classical position and momentum space. However, both random walks share an essential feature: they are sequences of deterministic evolutions separated by instantaneous random jumps. Therefore, the classical random walk techniques that have been developed with a high degree of sophistication should have some relevance for stochastic wave function approaches.

2.4 From quantum optics to classical random walks

A complete description of the random walk performed by the stochastic wave function in Hilbert space would be rather difficult and outside the scope of this work. We replace in the following such a full quantum analysis by a simpler description in terms of an inhomogeneous random walk performed by the momentum **p** of a fictitious classical particle. We keep in this classical description the important ingredients provided by the stochastic wave function approach, in order to be able to understand the basic features of the cooling process.

2.4.1 Fictitious classical particle associated with the quantum random walk

Let us come back to the quantum jump simulation of subrecoil cooling presented in Fig. 2.1. Such a random evolution looks like the random walk of a fictitious classical particle whose momentum p would change in a random way at certain times. It is clear that the evolution of such a classical particle cannot fully represent the quantum evolution of an ultracold atom in a subrecoil cooling experiment. Between two successive jumps, the state of the fictitious classical particle is fully characterized by a single number p, whereas the state of the ultracold atom is described by a wave function. Nevertheless, one must not forget that the simulation of Fig. 2.1 is derived from a rigorous quantum procedure. There are two ingredients of this quantum evolution which can be extracted and incorporated in the classical random walk, with the expectation that they could lead to a correct description of the cooling process.

First, the distribution of the delays between two successive spontaneous emissions is exactly calculated from an effective Hamiltonian. One can thus, at least in principle, exactly describe the distribution of the time intervals between two successive jumps performed by the fictitious classical particle. In practice, a further simplification will be introduced, as explained in the next section, which consists of taking simpler mathematical forms for the jump rate.

Second, in the simulation of Fig. 2.1, p is a constant of the motion of the quantum system between two successive jumps (the quasi-momentum characterizing the subspace of the Hilbert space where the coherent evolution takes place, i.e. the total momentum of the {atom+laser photons} system). If we consider this constant of the motion as a physical quantity characterizing the fictitious classical particle, it will remain constant, as it should, between two 'collisions' experienced by such a particle.

Using the delay function and the constants of the motion of the quantum problem, we can thus introduce a classical random walk which reproduces a certain number of important features of the quantum random walk.

2.4.2 Simplified jump rate

The most important feature of the quantum jump simulation of Fig. 2.1 is the appearance of very long waiting times τ between two successive jumps when $\mathbf{p} \to \mathbf{0}$. We will thus take for the inhomogeneous random walk of the corresponding classical particle an average jump rate $R(\mathbf{p})$, which exactly vanishes for $\mathbf{p} = \mathbf{0}$ and which has a behaviour around this point characterized by the exponent α of the lowest-order term in the expansion of $R(\mathbf{p})$ in powers of $\mathbf{p}$. These features are extracted from the properties of the delay function $W_\mathbf{p}(\tau)$ which has a long tail in τ when $\mathbf{p} \to \mathbf{0}$, the corresponding decay rate decreasing as p^α when $p = \|\mathbf{p}\| \to 0$.

Strictly speaking, $W_\mathbf{p}(\tau)$ is the modulus of a sum of complex exponentials of τ. Our simplification consists of keeping only the exponential with the longest time constant, which tends to ∞ when $p \to 0$. Thus $W_\mathbf{p}(\tau)$ can be simply described by a jump rate $R(\mathbf{p})$:

$$W_\mathbf{p}(\tau) \simeq R(\mathbf{p})e^{-R(\mathbf{p})\tau}. \tag{2.6}$$

This approximation can be justified by the following argument. The important point in the cooling schemes described here is the fact that the quantum system arriving after a jump in a family $\mathcal{F}(\mathbf{p})$ with $\mathbf{p}$ close to 0 has a non-zero probability to remain there for a very long time (which tends to ∞ when $p \to 0$). It could eventually make a jump after a very short time, because $W_\mathbf{p}(\tau)$ also contains rapidly decaying exponential components but, after a certain time, the system will come back in the neighbourhood of $\mathbf{p} = \mathbf{0}$, one or several times, until it stays there for a very long time. Such very long sojourn times are the origin of the 'anomalous' character of the random simulation of Fig. 2.1. By keeping only the longest time constant in $W_\mathbf{p}(\tau)$, one can hope to keep this essential ingredient which will allow us to derive correctly the asymptotic properties of the cooled atoms. Including the other shorter time constants would change only some prefactors[3], but would not modify the asymptotic θ-dependence of the various physical quantities.

Similarly, we will characterize the classical random walk far from $\mathbf{p} = \mathbf{0}$ by the behaviour of $R(\mathbf{p})$ for larger values of $\mathbf{p}$. We will introduce in Chapter 3 simple

---

[3] See Appendix A, in particular eq. (A.35), where we derive these prefactors for one-dimensional VSCPT.

models corresponding to different possible physical situations. Here also, we will find that there are situations leading to anomalous random walks.

2.4.3 Discussion

The basic idea of this simplified model is that the efficiency of subrecoil cooling is linked to the slowing down of the random walk around $\mathbf{p} = \mathbf{0}$, but that details of the exact characteristics of the random walk are unimportant. We are in fact following the usual approach in statistical physics, where general and powerful results can be found independently of microscopic details, provided that some essential features are taken into account. It seems difficult to demonstrate rigorously the validity of this approach, but of course we will compare its results to the results of the quantum microscopic calculations in the cases where such results are available, either from a GOBE treatment, or from a quantum jump treatment.

A benefit of our statistical approach is to yield quantitative predictions, even in cases where the quantum microscopic treatment is unable to make such predictions. It will thus enable us to address a few important questions such as the asymptotic behaviour at very long interaction times ($\theta \rightarrow \infty$) or the efficiency of subrecoil cooling in a configuration of dimension D larger than one.

3

Trapping and recycling. Statistical properties

In this chapter, we introduce two basic statistical distributions suited to an analysis of the classical inhomogeneous random walk that we introduced at the end of Chapter 2 for modelling non-ergodic cooling. These two distributions will be used throughout the book for deriving physically relevant quantities. The fact that they can be broad, with power-law tails, will also demonstrate from the beginning that Lévy statistics is naturally involved in non-ergodic cooling.

We begin in Section 3.1 by describing the evolution of the atom as a sequence of *trapping* processes of duration τ alternating with *recycling* processes of duration $\hat{\tau}$. This description will yield both physical insight and convenient calculations provided only two probability distributions are known, the distribution $P(\tau)$ of trapping times and the distribution $\hat{P}(\hat{\tau})$ of recycling times. In order to derive $P(\tau)$ and $\hat{P}(\hat{\tau})$, we then introduce in Section 3.2 simple physical models of the inhomogeneous jump rate. We then calculate $P(\tau)$ in Section 3.3 and $\hat{P}(\hat{\tau})$ in Section 3.4, using random walk techniques.

3.1 Trapping and recycling regions

As explained at the end of Chapter 2, we replace the microscopic quantum description of the evolution of the atom by a simpler description, where we consider a fictitious classical particle, completely characterized by its momentum **p**, and making a random walk with a step of rms length Δp of the order of $\hbar k$. This random walk takes place in a space that can have any dimension $D = 1, 2, 3$. Subrecoil cooling is characterized by an *inhomogeneous* jump rate, depending on the position **p** in the momentum space.

The observation of the random walks of individual atoms (Fig. 2.1) suggests distinguishing two regions in momentum space, a 'trapping region' around $p = 0$ and a 'recycling region' far from $p = 0$. Indeed, when an atom reaches $p \simeq 0$ states, it can remain 'trapped' for a relatively long time. In some cases it stays

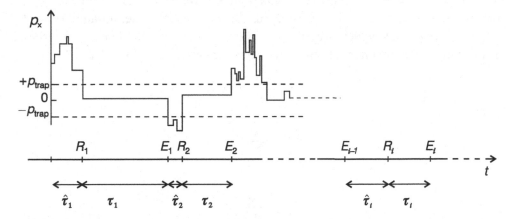

Fig. 3.1. Trapping times τ_i and recycling times $\hat{\tau}_i$. The atom returns to the trap at times $R_1, R_2, \ldots$, and exits the trap at times $E_1, E_2, \ldots$ (see Section 5.1.2).

there till the end of the laser–atom interaction. In other cases it scatters a photon before the laser is switched off which usually kicks it away from the $p \simeq 0$ region. Therefore, the atom will scatter photons again undergoing a random walk in p-space. This random walk will eventually lead the atom back to the trapping region again. Thus the atoms being kicked out of the trapping region are not lost, they are rather 'recycled' since the random walk process gives them other opportunities to reach long-lived small-p states. We introduce a momentum trap size p_{trap} to separate the two regions

$$\text{trapping region: } p \leq p_{\text{trap}}, \tag{3.1a}$$

$$\text{recycling region: } p \geq p_{\text{trap}}. \tag{3.1b}$$

The trap size p_{trap} is in principle arbitrary. We will indeed see that the physical observables no longer depend on p_{trap} in the limit $\theta \to \infty$. This trap size will be chosen conveniently below to simplify further calculations. In particular, p_{trap} will be taken to be smaller than the width p_0 of the jump rate dip (see Section 3.2):

$$p_{\text{trap}} < p_0. \tag{3.2}$$

The evolution of each atom now appears as a sequence of trapping periods of durations $\tau_1, \tau_2, \tau_3, \ldots$ alternating with recycling periods of durations $\hat{\tau}_1, \hat{\tau}_2, \hat{\tau}_3, \ldots$ (see Fig. 3.1). The $\hat{\tau}_i$'s are usually called 'first return times'.

During the interaction time θ, an atom is trapped N times, N being possibly different for each atom. During θ, the same atom has therefore been recycled also N times or $N \pm 1$ times depending on whether the atom was initially (and finally) in the trap. As we will be interested in long times θ, we have $N \gg 1$ and therefore, we consider $N \simeq N \pm 1$. If one disregards the last event[1] (either a trapping event or a recycling event) which overlaps the time $t = \theta$, the interaction time θ writes as the sum of the total trapping time T_N and the total recycling time $\hat{T}_N$:

$$\theta \simeq T_N + \hat{T}_N, \tag{3.3}$$

with

$$T_N = \sum_{i=1}^{N} \tau_i, \tag{3.4a}$$

$$\hat{T}_N = \sum_{i=1}^{N} \hat{\tau}_i. \tag{3.4b}$$

The sum T_N is the *total trapping time*, whereas $\hat{T}_N$ is the *total recycling time*, for an interaction time θ.

Both the τ_i's and the $\hat{\tau}_i$'s are independent random variables. Therefore, to study the statistical properties of the sums T_N and $\hat{T}_N$, one can think of using Central Limit Theorems (CLTs): from the probability distributions $P(\tau)$ (or $\hat{P}(\hat{\tau})$) of individual events, one infers the probability distribution $P_N(T_N)$ (or $\hat{P}_N(\hat{T}_N)$) of the sums.

A key point of the present work is that $P(\tau)$ is in many cases a 'broad' distribution, i.e. a distribution decaying so slowly at large τ that the second moment $\langle \tau^2 \rangle$ and even the first moment $\langle \tau \rangle$ are formally infinite (the same is true of $\hat{P}(\hat{\tau})$). This could be suspected from the graphical aspects of the Monte Carlo random walks of Fig. 2.1 which tend to generate very long trapping times. Usually, the finiteness of the first two moments ensures, via the CLT, that the sums T_N are distributed according to Gaussian laws ('normal' distributions) for large N. Here, the usual CLT is not directly applicable since $\langle \tau^2 \rangle$ (or even $\langle \tau \rangle$) diverges. On the other hand, if $P(\tau)$ behaves as a power law, $\tau^{-(1+\mu)}$ for large τ (which is the case here), one can use the generalized CLT of Lévy and Gnedenko. The distributions $P_N(T_N)$ no longer tend to normal distributions at large N but rather to 'Lévy distributions'.

As will be discussed below, Lévy distributions differ qualitatively from the normal distribution. For this reason, the appearance of Lévy statistics in subrecoil cooling has dramatic physical consequences. Indeed, the divergence of the average trapping time will be shown to be deeply related to the main features of the cooling mechanism, such as non-ergodicity. Thus non-ergodic cooling will appear to be qualitatively different from cooling with friction forces. In order to carry out

---

[1] This last event will be given a correct treatment in the quantitative calculations presented in subsequent chapters.

precise calculations, one needs to derive first the distributions $P(\tau)$ and $\hat{P}(\hat{\tau})$ of elementary events. This requires modelling of the inhomogeneous momentum random walk.

3.2 Models of inhomogeneous random walks

The distributions $P(\tau)$ and $\hat{P}(\hat{\tau})$ are determined by the random walk in momentum space. This random walk itself depends on the inhomogeneous jump rate $R(\mathbf{p})$ and on the possible existence of friction forces. In this section, we will introduce three models of inhomogeneous random walks that share the same features in the trapping region and that differ only in the recycling region. Note that we consider the random walk to be isotropic. Therefore, the jump rate $R(\mathbf{p})$ depends only on the atomic momentum modulus $p = \|\mathbf{p}\|$.

3.2.1 Friction

Before discussing both regions, we need a simple description of the friction forces that might be present. In usual laser cooling, the cooling effect of friction forces combined with the heating effect of spontaneous emission generates an approximately Gaussian stationary momentum distribution of half-width $p_{\max}$. In optimized low-intensity Doppler cooling, for instance, one has $p_{\max} \simeq (M\hbar\Gamma)^{1/2}$ where Γ^{-1} is the lifetime of the excited state. The friction forces vanish for $p \rightarrow 0$. Moreover the Gaussian momentum distribution decays very rapidly for $p > p_{\max}$. Therefore, it is reasonable to model friction forces very simply by a perfect 'wall' at $p_{\max}$ in momentum space. For $p < p_{\max}$, we consider that the atoms diffuse freely as if there was no friction, but no atomic momentum is allowed to be larger than $p_{\max}$. In other words, the real random walk with friction that explores in principle all the momentum space is, for our purposes, efficiently modelled by a standard (frictionless) random walk *confined* to a sphere of radius $p_{\max}$.

3.2.2 Trapping region

We can now establish the modelling of $R(p)$ in the trapping region, i.e. in the vicinity of $p = 0$. In all cases of non-ergodic cooling, $R(p)$ presents a dip of width p_0 around $p = 0$ which we assume to behave as a power law:

$$R(p) = \frac{1}{\tau_0}\left(\frac{p}{p_0}\right)^{\alpha}, \qquad p < p_0. \qquad (3.5)$$

The case of VSCPT corresponds to $\alpha = 2$ [AAK89] (see Appendix A). The flexibility of Raman cooling [KaC92] allows, in principle, any value of α. Up to now, Raman cooling experiments have used configurations with $\alpha = 4$ and $\alpha = 2$ [RBB95]. Note that the friction forces which might be present are assumed to play no role for $p < p_0$ as p_0 will always be taken smaller than p_{max}. Although the function $R(p)$ of eq. (3.5) obviously depends on a single parameter $\tau_0 p_0^\alpha$, we have introduced the value p_0 of the momentum at which the jump rate $R(p)$ saturates and takes the value τ_0^{-1} (Fig. 3.2). The parameter p_0 then characterizes the width of the dip of $R(p)$ around $p = 0$, while τ_0^{-1} is the jump rate at saturation.

Spurious mechanisms can cause the cancellation of $R(p)$ at $p = 0$ to be imperfect. In these cases, eq. (3.5) must be replaced by

$$R(p) = R_0 + \frac{1}{\tau_0}\left(\frac{p}{p_0}\right)^\alpha, \quad p < p_0. \tag{3.6}$$

In most of this book we will only consider that $R_0 = 0$. The cases $R_0 > 0$, which can be important for practical applications, can easily be taken into account with our approach. This is done in Section 7.4.

3.2.3 Recycling region

Consider now the possible models for the recycling region, i.e. the region $p > p_0$ out of the dip. In the first model, the jump rate is assumed to be constant for all $p > p_0$ and the atomic momentum is confined to a sphere of radius p_{max}:

$$\text{confined model:} \quad R(p) = \frac{1}{\tau_0}, \quad p_0 \le p \le p_{max}. \tag{3.7}$$

This *confined* model describes faithfully most situations of friction-assisted subrecoil cooling [MaA91, SHP93, WEO94, MDT94, LBS94, LKS95, HLO00].

In the second model, the jump rate is also assumed to be constant for all $p > p_0$ but the atomic momentum random walk is allowed to go to infinity ($p_{max} \to \infty$):

$$\text{unconfined model:} \quad R(p) = \frac{1}{\tau_0}, \quad p_0 \le p. \tag{3.8}$$

This *unconfined* model is well suited to cases in which the atomic momentum diffusion is frictionless and when the optical Doppler effect, which shifts the atoms out of resonance with the cooling lasers at large p, can be neglected.

The third model assumes unconfined momentum diffusion and takes into account the decrease of the jump rate due to the *Doppler* effect:

$$\text{Doppler model:} \quad R(p) = \frac{1}{\tau_0}, \quad p_0 \le p \le p_D, \tag{3.9a}$$

$$R(p) = \frac{1}{\tau_0}\left(\frac{p_D}{p}\right)^2, \quad p_D \le p, \tag{3.9b}$$

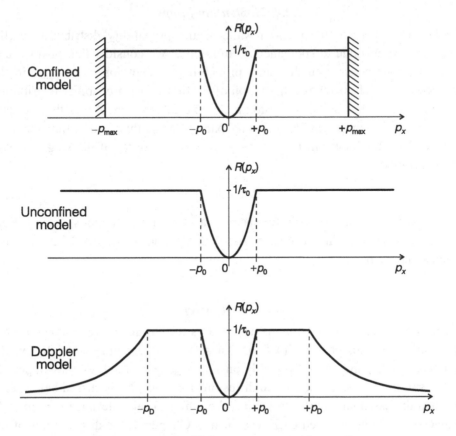

Fig. 3.2. Models of inhomogeneous random walks.

where the characteristic momentum p_D is defined by $k p_D / M = \Gamma/2$ (k is the laser wave number). The previous equations for $R(p)$ are obtained by taking the small-p and the large-p limits of the Lorentzian $\Gamma^2/(\Gamma^2 + 4k^2 p^2/M^2)$, describing the decrease of the jump rate due to the Doppler shift kp/M (see Appendix A, p. 152). This Doppler model describes faithfully the original one-dimensional VSCPT scheme with σ_+/σ_- polarization [AAK88].

A fourth model could be introduced, with *absorbing walls* at $p = p_{abs}$. These absorbing walls would account for momentum dependent forces that appear for $p > p_{max}$ in some experiments [LBS94] that tend to push the atoms towards *larger* momenta, unlike friction forces. This undesirable effect can in principle be reduced by an adequate choice of experimental parameters. It will therefore not be studied here.

3.2.4 Momentum jumps

The last ingredient of the random walks is the probability distribution of the momentum jumps due to spontaneous emissions. We consider that positive and negative jumps occur with the same probability (except for $p = p_{max}$ in the confined model, cf. the above discussion of friction). The probability distribution of jumps spans an interval of approximate size $2\Delta p$ where Δp is the standard deviation of the jump lengths, the only parameter of this distribution that will be needed in this book. In most cases, Δp is of the order of the single photon momentum $\hbar k$:

$$\Delta p \simeq \hbar k. \tag{3.10}$$

One can calculate Δp precisely for specific laser cooling situations (see e.g. Section A.1.2.6, for one-dimensional σ_+/σ_- VSCPT and Section A.2.2.3, for one-dimensional Raman cooling).

3.2.5 Discussion

To sum up, physical considerations have led us to introduce three models for the inhomogeneous momentum diffusion. These models depend essentially on the parameters α, p_0 and τ_0, and possibly on R_0, p_{max} and p_D. They may appear to be oversimplifications of the atomic diffusion. However, as discussed below, they do grasp the essential features of subrecoil cooling. Their relevance is intimately connected to the generalized CLT: as shown in Chapter 4, the distributions of the sums T_N and $\hat{T}_N$ at large N depend *only* on the asymptotic behaviour of $P(\tau)$ and $\hat{P}(\hat{\tau})$ when these distributions are broad. So the only requirement on the models for predictions in the long time regime (large N) is that they describe correctly the *asymptotic* behaviours of $P(\tau)$ and $\hat{P}(\hat{\tau})$. Therefore, these simplified models will allow *exact* analytical predictions in the long time limit.

3.3 Probability distribution of the trapping times

3.3.1 One-dimensional quadratic jump rate

We first consider the case of a one-dimensional random walk along the p_x axis, with a quadratic variation of the jump rate around $p_x = 0$:

$$R(p_x) = \frac{1}{\tau_0}\left(\frac{p_x}{p_0}\right)^2 = \frac{1}{\tau_0}\left(\frac{p}{p_0}\right)^2. \tag{3.11}$$

Bearing in mind that the real motion takes place in a three-dimensional space, the lengths of the steps of the random walk are not all equal, but rather random between $\simeq -\Delta p$ and $\simeq +\Delta p$, corresponding to the projection of the recoil momentum (of random direction) onto the x axis. All the points of the p_x axis can then be explored by the diffusing atom. If we now make the further assumption that the trap size p_{trap} is small compared to the step length

$$p_{trap} \ll \Delta p, \tag{3.12}$$

then all the points in the trapping region are reached with the same probability ('uniform sprinkling'). The probability density $\rho(p_x)$ that an atom entering the trap of width $2 p_{trap}$ reaches the momentum p_x is therefore approximated as:

$$\rho(p_x) = \frac{1}{2\,p_{trap}}. \tag{3.13}$$

The probability for a trapped atom making a momentum jump to fall back into the trap is of the order of $p_{trap}/\Delta p$, which is negligibly small because of the inequality (3.12). The trapping time $\tau(p_x)$ for an atom landing in the trap at p_x is therefore equal to the time spent at p_x, which is directly related to the jump rate $R(p_x)$ of eq. (3.11). In other words, the jump rate $R(p_x)$ is also the *rate of escape* from the trap, for the atoms with $|p_x| < p_{trap}$.

3.3.1.1 Deterministic model

Let us first assume, for simplicity, that an atom entering the trap with a momentum p_x remains there for a well defined, deterministically fixed, time $\tau(p_x)$ given by:

$$\tau(p_x) = \frac{1}{R(p_x)} = \tau_0 \left(\frac{p_0}{p_x} \right)^2 \tag{3.14}$$

(in reality, the time τ is itself random, distributed according to an exponential law of mean $1/R(p_x)$; we shall take this into account below, see eq. (3.19)). The trapping times $\tau(p_x)$ are therefore distributed (Fig. 3.3) between a minimum value τ_{trap} (corresponding to $p_x = \pm p_{trap}$)

$$\tau_{trap} = \tau(p_{trap}) = \tau_0 \left(\frac{p_0}{p_{trap}} \right)^2 \tag{3.15}$$

and infinity (corresponding to $p_x = 0$), with a probability distribution $P(\tau)$ such that

$$P(\tau)\,|d\tau| = 2\rho(p_x)\,|dp_x| \tag{3.16}$$

which means that all events either between p_x and $p_x + dp_x$, or between $-p_x$ and $-p_x - dp_x$, contribute to trapping times between τ and $\tau + d\tau$, where $|d\tau|$ and $|dp_x|$ are related by the equation

$$\frac{|d\tau|}{|dp_x|} = |\tau'(p_x)| = 2\tau_0 \frac{p_0^2}{|p_x|^3} \tag{3.17}$$

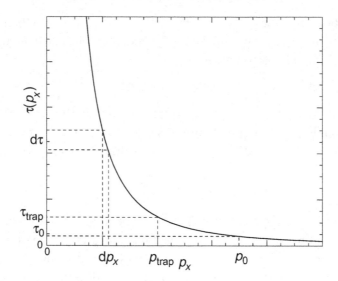

Fig. 3.3. Trapping times $\tau(p_x)$ for $p_x > 0$.

(see eq. (3.14) and Fig. 3.3). Inserting eq. (3.13) and eq. (3.17) into eq. (3.16), we then find the distribution of the trapping times to be

$$P(\tau) = \frac{\tau_{\text{trap}}^{1/2}}{2\tau^{3/2}}, \quad \tau \geq \tau_{\text{trap}}, \tag{3.18}$$

with τ_{trap} defined by eq. (3.15).

This probability distribution, shown in Fig. 3.4, is a broad function with slowly decreasing tails. The probability of observing large values of τ is so important that the average value of τ is infinite. This unusual behaviour is precisely at the root of the efficiency of the subrecoil cooling mechanisms, which are based on the existence of very long trapping times around $p_x = 0$.

3.3.1.2 Exponential model

The $\tau^{-3/2}$ behaviour of the distribution of the trapping times for large τ is the main result of the above calculation. This result is not substantially modified if one considers a more realistic model, where the trapping time for a given momentum is an exponential random variable rather than a deterministic variable. If $R(p_x)$ is the jump rate associated with a trapped momentum p_x, the conditional distribution of the trapping times for a well defined p_x is

$$P(\tau \mid p_x) = R(p_x) \exp\left(-R(p_x)\,\tau\right). \tag{3.19}$$

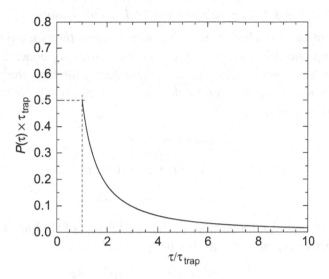

Fig. 3.4. Distribution $P(\tau)$ (deterministic model).

The total probability distribution of the trapping times is then

$$P(\tau) = \int_{-p_{\text{trap}}}^{+p_{\text{trap}}} P(\tau \mid p_x)\, \rho(p_x)\, dp_x. \tag{3.20}$$

Using the expression for the jump rate (eq. (3.11)) and the uniform distribution for entering the trap at momentum p_x (eq. (3.13)), one finds, after changing variables to $u = p_x^2 \tau/(p_0^2 \tau_0)$:

$$P(\tau) = \frac{1}{2}\frac{\tau_{\text{trap}}^{1/2}}{\tau^{3/2}}\, \gamma\left(1 + \frac{1}{2}, \frac{\tau}{\tau_{\text{trap}}}\right), \tag{3.21}$$

where $\gamma(\beta, x)$ is the incomplete Gamma function defined by

$$\gamma(\beta, x) = \int_0^x e^{-u} u^{\beta-1}\, du. \tag{3.22}$$

Taking the limit $\tau \to \infty$, one has

$$\gamma\left(1 + \frac{1}{2}, \frac{\tau}{\tau_{\text{trap}}}\right) \to \Gamma\left(1 + \frac{1}{2}\right) = \frac{1}{2}\Gamma\left(\frac{1}{2}\right).$$

The asymptotic behaviour of $P(\tau)$ is thus given by:

$$P(\tau) \underset{\tau \gg \tau_{\text{trap}}}{\simeq} \frac{\Gamma(1/2)}{2}\frac{\tau_{\text{trap}}^{1/2}}{2\tau^{3/2}} \tag{3.23}$$

where $\Gamma(1/2) = \sqrt{\pi}$. For finite values of τ, there are subleading correction terms which can be systematically calculated (see eq. (6.5.32) in [AbS70]).

3.3.2 Generalization to higher dimensions

If we now consider a random walk in a D-dimensional momentum space, with a quadratic jump rate still given by eq. (3.11), we can easily generalize the above results. Let us again assume that, for an atom entering the trap, the probability to land anywhere in the trapping volume $V_D(p_{trap})$ is uniform. The volume $V_D(p)$ of the hypersphere of radius p reads:

$$V_D(p) = C_D p^D, \tag{3.24}$$

where C_D is the volume of the unit sphere in D dimensions:

$$C_1 = 2, \quad C_2 = \pi, \quad C_3 = 4\pi/3. \tag{3.25}$$

The probability $\rho(p)\,dp$, for an atom landing in the trap, of landing at a momentum of modulus between p and $p + dp$ is simply given by:

$$\rho(p)dp = \frac{dV_D(p)}{V_D(p)} = \frac{S_D p^{D-1}\,dp}{V_D(p_{trap})}, \tag{3.26}$$

where $S_D p^{D-1}$ is the surface of a hypersphere of radius p ($S_D = DC_D$):

$$S_1 = 2, \quad S_2 = 2\pi, \quad S_3 = 4\pi. \tag{3.27}$$

Thus, one obtains

$$\rho(p) = \frac{Dp^{D-1}}{p_{trap}^D}. \tag{3.28}$$

Calculations similar to those of the one-dimensional case then lead to

$$P(\tau) \underset{\tau \gg \tau_{trap}}{\simeq} \mathcal{A}\frac{D\tau_{trap}^{D/2}}{2\tau^{1+D/2}}, \tag{3.29}$$

where $\mathcal{A}$ is a numerical factor which depends on whether one assumes a deterministic (see eq. (3.14)) or an exponential (see eq. (3.19)) relation between $\tau(p)$ and $R(p)$. In the former case, $\mathcal{A} = 1$, while in the latter case, $\mathcal{A} = (D/2)\,\Gamma(D/2)$. Note that as soon as $D > 2$, the average trapping time is finite (although the variance of the trapping time still diverges if $D < 4$).

3.3.3 Generalization to a non-quadratic jump rate

If we consider more general situations where the jump rate varies as p^α (cf. eq. (3.5)), very similar calculations lead to the following result:

$$P(\tau) \underset{\tau \gg \tau_{trap}}{\simeq} \mathcal{A}_\mu \frac{\mu \tau_{trap}^\mu}{\tau^{1+\mu}} \quad \text{with} \quad \mu \equiv \frac{D}{\alpha}, \tag{3.30}$$

where the characteristic trapping time τ_{trap} is defined as:

$$\tau_{\text{trap}} = \tau_0 \left(\frac{p_0}{p_{\text{trap}}} \right)^\alpha . \qquad (3.31)$$

The numerical constant $\mathcal{A}_\mu$ is still equal to one in the deterministic case, and to $\mathcal{A}_\mu = \mu \Gamma(\mu)$ in the exponential case.

3.3.4 Discussion

We have thus shown that the asymptotic behaviour of the trapping time distribution $P(\tau)$ at large τ decays as a *power law* with an exponent μ given by the ratio of the dimension D of the momentum space to the exponent α characterizing the p-dependence of the jump rate $R(p)$ around the trapping point. This power-law distribution is conveniently written as

$$P(\tau) \underset{\tau \gg \tau_{\text{b}}}{\simeq} \frac{\mu \tau_{\text{b}}^\mu}{\tau^{1+\mu}} \qquad (3.32)$$

with

$$\mu \equiv \frac{D}{\alpha} \qquad \text{and} \qquad \tau_{\text{b}}^\mu \equiv \mathcal{A}_\mu \tau_{\text{trap}}^\mu = \mathcal{A}_\mu \left(\frac{p_0}{p_{\text{trap}}} \right)^D \tau_0^\mu, \qquad (3.33)$$

where $\mathcal{A}_\mu$ is defined by

$$\text{deterministic case:} \quad \mathcal{A}_\mu = 1, \qquad (3.34a)$$

$$\text{exponential case:} \quad \mathcal{A}_\mu = \mu \Gamma(\mu). \qquad (3.34b)$$

Note that, in the deterministic case, the expression (3.32) is exact for all $\tau \geq \tau_{\text{trap}}$.

When $\mu \leq 2$, the variance of τ does not exist, and the usual (Gaussian) CLT does not apply (see Chapter 4). When $\mu \leq 1$ the tails of the probability distribution are so broad that the average value of τ fails to converge. Such a situation is *a priori* favourable for efficient cooling, since it corresponds to the case where very long trapping times around $p = 0$ have a substantial probability.

In such situations, however, one cannot apply usual statistical treatments. In particular, when $\mu \leq 1$, the total trapping time T_N is clearly not proportional to $N\langle \tau \rangle$ (where N is the number of trapping events), since $\langle \tau \rangle$ is infinite. One thus has to resort to the generalized CLT of Lévy and Gnedenko (see Chapter 4). In the intermediate case $1 < \mu \leq 2$, the average trapping time is finite and differs from τ_{trap} by a factor which diverges as μ tends to one from above. For example, in the deterministic case, one gets:

$$\langle \tau \rangle = \frac{\mu}{\mu - 1} \tau_{\text{trap}}. \qquad (3.35)$$

In order to have μ as small as possible, which is *a priori* favourable for efficient cooling, the exponent α must be large: this corresponds to 'flat' behaviour of the jump rate $R(p)$ around the trapping point. On the other hand, when the number D of dimensions increases, the tails of $P(\tau)$ decay faster, simply because of the phase space relation (eq. (3.28)), which gives less weight to small values of p when the space dimension increases.

To make the connection with real subrecoil cooling schemes (Appendix A), notice that the initial VSCPT scheme in one dimension corresponds to a broad distribution where $\langle \tau \rangle$ is infinite ($\mu = 1/2$), while in the three-dimensional case ($\mu = 3/2$) the average of τ does exist. The two-dimensional situation corresponds to the marginal case $\mu = 1$. Raman cooling corresponds to $\alpha \simeq 4$ when using Blackman pulses and to $\alpha = 2$ when using square time pulses [RBB95].

3.4 Probability distribution of the recycling times

3.4.1 *Presentation of the problem: first return time in Brownian motion*

In contrast to a trapping period, which consists of a single event (the atom is trapped at a given **p**), a recycling period is a random walk composed of many steps out of the trap. We characterize such a composite recycling period by a single number $\hat{\tau}$, which is the recycling time, i.e. the time needed to return to the trapping region. The aim of this section is to establish the probability distribution $\hat{P}(\hat{\tau})$ of the recycling times $\hat{\tau}$. This is in fact a 'first return time' problem, a standard problem in Brownian motion theory: $\hat{\tau}$ can be identified as the time needed for the random walk in momentum space to return to the origin.

It is well known that this problem depends crucially on the dimension of the space where Brownian motion takes place. Indeed, in the one-dimensional case, the probability that a random walker returns to the origin is equal to one (actually, the random walker returns infinitely often to its starting point). On the contrary, in dimensions greater than $D = 2$, there is a non-zero probability that the walker never returns to its starting point.

Another important parameter controlling the first return time is the average duration $1/R(p)$ of the steps of the random walk. We calculate below the recycling time distributions $\hat{P}(\hat{\tau})$ for the three models introduced in Section 3.2. In the *unconfined model* (Section 3.4.2), $R(p)$ is constant outside the trap, and the motion is the usual random walk with a uniform jump rate. In the *Doppler model* (Section 3.4.3), $R(p)$ decreases for large values of p; recycling walks reaching large values of p are slowed down, and large recycling times are obviously more probable than in the unconfined model: we therefore expect a broader distribution for the recycling times $\hat{\tau}$. On the contrary, in the *confined model* (Section 3.4.4), the random walk motion is bounded by 'hard walls', and one expects that large recycling times are scarce, leading to a relatively narrow distribution $\hat{P}(\hat{\tau})$.

3.4.2 The unconfined model in one dimension

We first recall here a few results for the problem of an atom making a one-dimensional uniform random walk on the p_x axis (p_x is the algebraic position which takes values between $-\infty$ and $+\infty$, in contrast to its modulus $p = |p_x|$ which is always positive). The average value of the elementary step is zero (isotropic random motion) and its variance Δp^2 is independent of the position p_x. The average time between two successive steps is finite and equal to τ_0 (unconfined model). At time $t = 0$ the atom leaves the trapping region around the origin ($p \leq p_{\text{trap}}$ with $p_{\text{trap}} \ll \Delta p$). We want to determine the probability distribution $\hat{P}(\hat{\tau})$ of the time $t = \hat{\tau}$ at which the system returns *for the first time* to the trapping region.

Let us start by determining the probability distribution $P_1(n)$ of the number n of steps needed to return for the first time to the trap. At this stage, working only in terms of the *number* of steps, we deal with a purely geometric problem and the existence of long-lived trapping states for $p < p_0$ plays no role. In order to solve this problem, we introduce the probability $P_{\text{trap}}(n)$ that the atom is in the trap after n steps, *independently of the number of previous returns*. This probability is the integral over the trapping region ($-p_{\text{trap}} \leq p_x \leq p_{\text{trap}}$) of the probability density $P(p_x, n)$ of p_x after n steps. For a standard random walk, it is well known that after a large number n of steps, the distribution of p_x is Gaussian:

$$P(p_x, n) = \frac{1}{\sqrt{2\pi n \, \Delta p^2}} \exp\left(-\frac{p^2}{2 n \, \Delta p^2}\right). \tag{3.36}$$

Using the condition $p_{\text{trap}} \ll \Delta p$, one thus finds

$$P_{\text{trap}}(n) = \int_{-p_{\text{trap}}}^{p_{\text{trap}}} \mathrm{d}p_x P(p_x, n) = \frac{2 \, p_{\text{trap}}}{\sqrt{2\pi n} \, \Delta p}. \tag{3.37}$$

We now want to relate $P_{\text{trap}}(n)$ to the first return distribution $P_1(n)$. The atom can be in the trap after n steps, either for the first time (with probability $P_1(n)$), or because it was already in the trap after $n' < n$ steps (with probability $P_{\text{trap}}(n')$), left the trap at the step $n' + 1$ and then returned once more after $n - n'$ steps (with probability $P_1(n - n')$). All the possibilities are covered by allowing n' to vary between 1 and $n - 1$. Summing over n', we can therefore write an exact relation:

$$P_{\text{trap}}(n) = \delta_{n,0} + \sum_{n'=0}^{n} P_{\text{trap}}(n') \, P_1(n - n'), \tag{3.38}$$

where the Kronecker symbol $\delta_{n,0}$ accounts for the fact that the atom is in the trap for $n = 0$. We have extended the summation from $n' = 0$ to $n' = n$, taking into account the fact that $P_{\text{trap}}(n = 0) = 1$ and $P_1(n = 0) = 0$.

One then introduces two generating functions (discrete Laplace transforms), as:

$$\mathcal{L}_{\text{d}} P_{\text{trap}}(s) = \sum_{n=0}^{\infty} e^{-sn} P_{\text{trap}}(n) \tag{3.39}$$

and similarly for $\mathcal{L}_{\text{d}} P_1$. Multiplying eq. (3.38) by e^{-sn} and summing over n leads to:

$$\mathcal{L}_{\text{d}} P_{\text{trap}}(s) = 1 + \mathcal{L}_{\text{d}} P_{\text{trap}}(s)\, \mathcal{L}_{\text{d}} P_1(s) \tag{3.40}$$

or

$$\mathcal{L}_{\text{d}} P_1(s) = 1 - \frac{1}{\mathcal{L}_{\text{d}} P_{\text{trap}}(s)}. \tag{3.41}$$

We are interested in the long time behaviour of $\hat{P}(\hat{\tau})$. Since the average time between two steps is finite, it is obvious that the large $\hat{\tau}$ regime corresponds to a large number of steps n. The information about large $\hat{\tau}$ is thus contained in the small s behaviour of $\mathcal{L}_{\text{d}} P_1(s)$. For small s, the region $n < s^{-1}$ does not contribute to leading order and the discrete sums over n can be replaced by integrals (corresponding to usual Laplace transforms). This gives, using eq. (3.36):

$$\mathcal{L}_{\text{d}} P_{\text{trap}}(s) \underset{s \to 0}{\simeq} \int_0^{\infty} dn\, e^{-sn} P_{\text{trap}}(n)$$

$$= \frac{2 p_{\text{trap}}}{\sqrt{2\pi}\, \Delta p} \int_0^{\infty} dn\, \frac{e^{-sn}}{\sqrt{n}} = \frac{\sqrt{2} p_{\text{trap}}}{\Delta p}\, \frac{1}{\sqrt{s}} \tag{3.42}$$

(we have used $\Gamma(1/2) = \sqrt{\pi}$). It then follows from eq. (3.41) that

$$\mathcal{L}_{\text{d}} P_1(s) \underset{s \to 0}{\simeq} 1 - \frac{\Delta p}{\sqrt{2} p_{\text{trap}}}\, \sqrt{s}. \tag{3.43}$$

Note that $\mathcal{L}_{\text{d}} P_1(s = 0) = \sum_{n=0}^{\infty} P_1(n) = 1$, which means that the total probability of returning to the origin is equal to one. As discussed in the next chapter, the small s behaviour of $\mathcal{L}_{\text{d}} P_1(s)$ and the large n behaviour of $P_1(n)$ are linked. From eq. (4.1) and eq. (4.14) of Chapter 4, one can deduce that:

$$P_1(n) \underset{n \to \infty}{\simeq} \frac{1}{2\sqrt{2\pi}}\, \frac{\Delta p}{p_{\text{trap}}}\, \frac{1}{n^{3/2}}. \tag{3.44}$$

We can now come to the time variable $\hat{\tau}$. The probability density of returning to the trap for the first time at time $\hat{\tau}$ is related to the probability density $P_1(n)$ for the number of steps through:

$$\hat{P}(\hat{\tau}) = \sum_{n=0}^{\infty} P_1(n)P(\hat{\tau}|n), \qquad (3.45)$$

where $P(\hat{\tau}|n)$ is the probability density that the n steps have taken a time $\hat{\tau}$. Since the average time between jumps is finite, the law of large numbers ensures that one can replace, to leading order in the large $\hat{\tau}$ limit, $P(\hat{\tau}|n)$ by $\delta(\hat{\tau} - n\tau_0)$ (τ_0 is the average jump time).

Using eq. (3.44), we finally obtain:

$$\hat{P}(\hat{\tau}) = \frac{\hat{\tau}_{\mathrm{b}}^{1/2}}{2\hat{\tau}^{3/2}} \quad \text{with} \quad \hat{\tau}_{\mathrm{b}} = \frac{1}{2\pi}\left(\frac{\Delta p}{p_{\mathrm{trap}}}\right)^2 \tau_0. \qquad (3.46)$$

This result, valid for large values of $\hat{\tau}$, shows that the recycling time distribution is very broad, with tails decreasing so slowly that the average value of the recycling times diverges. This calls for the use of Lévy statistics, which we shall introduce in the next chapter.

The presence of the ratio $\Delta p/p_{\mathrm{trap}}$ in expression (3.46) has an interesting interpretation. When an atom, making steps of typical size Δp, comes back in the vicinity of $p = 0$, it has a probability $p_{\mathrm{trap}}/\Delta p$ of falling into the trap. Therefore, this atom must come back typically $\Delta p/p_{\mathrm{trap}}$ times *in the vicinity of* $p = 0$ in order to have an appreciable probability to return *to the trap*. The larger Δp compared to p_{trap}, the higher the probability to 'miss' the trap when coming back to the vicinity of $p = 0$, and therefore the larger the typical return time $\hat{\tau}_{\mathrm{b}}$ to the trap.

Furthermore, the power 2 of $(\Delta p/p_{\mathrm{trap}})^2$ can also be easily understood. It comes from the fact that the mth first return path ($m = \Delta p/p_{\mathrm{trap}}$) is typically m^2 times longer than the first return path (see the properties of Lévy sums, Section 4.3.1).

The return time distribution becomes even broader for higher dimensions, where the atom on a random walk has difficulty relocating its initial site. In two dimensions, $\hat{P}(\hat{\tau})$ only decays as $\hat{\tau}^{-1}\log^{-2}(\hat{\tau})$, whereas in three dimensions, there is a finite probability that the walk never returns, which corresponds to a non-zero weight of $\hat{P}(\hat{\tau})$ at $\hat{\tau} = \infty$ [Wei94].

3.4.3 The Doppler model in one dimension

We now consider the case where the jump rate $R(p)$ decreases for large values of p. We bear in mind the experiments of frictionless one-dimensional VSCPT, where the rate of fluorescence decreases as a consequence of the Doppler shift. We thus specifically take the case described by eq. (3.9), corresponding to the Lorentzian wing of the atomic fluorescence.

An exact calculation of the tail of the probability distribution $\hat{P}(\hat{\tau})$ actually turns out to be possible in this case, and is presented in Appendix B. Only a simplified argument, which reproduces the correct form of this tail, is given here.

We first notice that the probability distribution of the number of first return steps $P_1(n)$ is a purely geometrical property, independent of the duration of each step, so that the expression eq. (3.44) is still valid. The proportionality between the return *time* and the *number* of steps is, however, no longer valid. During an n steps long walk, the typical distance p_n covered by the walk is $\Delta p \sqrt{n}$, each small interval of size dp being visited typically $n dp/(\Delta p \sqrt{n})$ times. The total time spent by the walker outside the trap can thus be approximated as:

$$\hat{\tau}(n) = \sum_{n'=1}^{n} \frac{1}{R(p_{n'})} \simeq \frac{dp\sqrt{n}}{\Delta p} \sum_{i=1}^{\Delta p\sqrt{n}/dp} \frac{1}{R(p_i = i\,dp)} \tag{3.47}$$

since each small interval of size dp will contribute $\sqrt{n}\,dp/\Delta p$ times. In the small dp limit, the sum can be replaced by an integral, and one finds using eq. (3.9):

$$\hat{\tau}(n) \simeq \frac{\sqrt{n}}{\Delta p} \tau_0 \int_0^{\Delta p\sqrt{n}} \frac{p^2}{p_D^2}\,dp \simeq \frac{\tau_0}{3} \left(\frac{\Delta p}{p_D}\right)^2 n^2. \tag{3.48}$$

Note that we take the expression (3.9) for $R(p)$ even when $p_0 < p < p_D$ since this region contributes negligibly to long $\hat{\tau}$'s and that the lower bound of the integral is safely extended to zero because the integral is dominated by large p's.

Finally, using the distribution $P_1(n)$ of the number of first return steps given by eq. (3.44), we obtain the distribution of the first return times as

$$\hat{P}(\hat{\tau}) = P_1(n)\frac{dn}{d\hat{\tau}} \simeq \frac{\hat{\tau}_b^{1/4}}{4\hat{\tau}^{5/4}} \tag{3.49}$$

with now, up to prefactors of the order of one which we calculate in Appendix B,

$$\hat{\tau}_b \simeq \tau_0 \frac{\Delta p^6}{p_{\text{trap}}^4\, p_D^2}. \tag{3.50}$$

This result shows that the distribution of recycling times has very broad tails, decaying as $\hat{\tau}^{-5/4}$, i.e. still more slowly than in the case of a uniform one-dimensional random walk. This is not surprising, since when the number of steps increases (so that $\hat{\tau}$ also increases), the jump rate slows down because it explores larger values of p where the Doppler effect plays a more important role. In this case also, the average return time is infinite.

For intermediate times, the Doppler effect can be neglected and the relevant jump rate is nearly constant. Therefore the $\hat{\tau}^{-3/2}$ law describes the return time distribution for times small compared to the diffusion time associated with p_D (see Section A.1.1.5, p. 153, and Section 8.3.2).

3.4.4 The confined model: random walk with walls

We now consider the *confined model*, where the random walk in a D-dimensional space is confined by reflecting walls on a sphere $p = p_{max}$. As for the other models, we first begin by reasoning only on the number n of steps, regardless of the time they take.

Since the motion is confined, the walk explores the sphere in a uniform way at large n. For large n, the probability of finding the atom in the trapping volume after n steps is thus simply equal to

$$P_{trap}(n) = \left(\frac{p_{trap}}{p_{max}} \right)^D, \tag{3.51}$$

i.e. the ratio of the trapping volume to the volume of the total space. The discrete Laplace transform of this function is:

$$\mathcal{L}_d P_{trap}(s) = \left(\frac{p_{trap}}{p_{max}} \right)^D \frac{1}{s}. \tag{3.52}$$

We can then obtain the probability distribution $P_1(n)$ of the number of steps for the first return times by using eq. (3.41), which is valid for all models. The Laplace transform of $P_1(n)$ is thus

$$\mathcal{L}_d P_1(s) = 1 - \left(\frac{p_{max}}{p_{trap}} \right)^D s. \tag{3.53}$$

The fact that the small s expansion of $\mathcal{L}_d P_1(s)$ starts with a term linear in s indicates that the average number of steps needed to return to the origin is finite, and is simply equal to the coefficient of s (see eq. (4.18)):

$$\langle n \rangle = \left(\frac{p_{max}}{p_{trap}} \right)^D. \tag{3.54}$$

We can now come to time variables. The average time τ_0 between two successive steps being finite, the average first return time $\hat{\tau}$ is now also finite (at variance with the *unconfined* and *Doppler* models), with:

$$\langle \hat{\tau} \rangle = \langle n \rangle \tau_0. \tag{3.55}$$

One thus finds:

$$\langle \hat{\tau} \rangle = \tau_0 \left(\frac{p_{max}}{p_{trap}} \right)^D. \tag{3.56}$$

This result is important, since many experiments are carried out in a situation where diffusion out of the trap is limited by a friction mechanism. It is thus worth some further comments.

- One could actually show [Wei94] that the full distribution $\hat{P}(\hat{\tau})$ decays exponentially for large $\hat{\tau}$, as $\exp(-\hat{\tau}/\tau_{max})$, where τ_{max} is the time needed for the random walk to reach the wall:

$$\tau_{max} \simeq \tau_0 \left(\frac{p_{max}}{\Delta p}\right)^2. \tag{3.57}$$

Correspondingly, the result (3.56) is valid when the time is large enough so that the system can explore all the accessible space, i.e. when the evolution time is much larger than τ_{max}. In the opposite limit, the results obtained for the *unconfined model* remain valid.

- Notice that $\langle\hat{\tau}\rangle$ increases very quickly with the dimension of the space when $p_{max} \gg p_{trap}$.

- It may appear surprising that the mean return time $\langle\hat{\tau}\rangle$ does not depend on the length Δp of the steps in the momentum space, at variance with the corresponding results for $\hat{\tau}_b$ in the unconfined model (cf. eq. (3.46)) and in the Doppler model (cf. eq. (3.50)). One can interpret this result by noting that when the length Δp of the individual step increases, the atom comes back faster close to the origin, but the probability of missing the trap increases because the sampling of space is coarser.

- There are several hidden assumptions in the above calculation, in particular when we have identified the average time between jumps with τ_0. This is not obvious when $p_{trap} \ll p_0$, since some jumps take place in the region where the jump rate has already substantially dropped. One can show that if $\Delta p \gg p_{trap}$, the average time between jumps remains of the order τ_0 for $D > \alpha$, while for $D < \alpha$, it is modified to:

$$\langle\tau\rangle \simeq \tau_0 \left(1 + \frac{D}{\alpha - D}\frac{p_{trap}^{D-\alpha}p_0^{\alpha}}{p_{max}^{D}}\right). \tag{3.58}$$

3.4.5 Discussion

In this chapter, we have established some results on the statistical properties of the recycling time $\hat{\tau}$, i.e. the delay between two successive trapping periods. If the random walk out of the trap is *confined* (corresponding to a realistic situation with friction, favourable to the cooling mechanism), then the average recycling time $\langle\hat{\tau}\rangle$ is finite. For N sequences of trapping and recycling, the total time $\hat{T}_N$ spent out of the trap is then given, at large N, by the usual law of large numbers:

$$\hat{T}_N \simeq N\langle\hat{\tau}\rangle. \tag{3.59}$$

On the contrary, if the random walk is not confined, the distribution of the recycling times is so broad that the average of $\hat{\tau}$ does not exist, and one cannot write an equation such as eq. (3.59). It will be possible, however, to determine the statistical properties of the total recycling time $\hat{T}_N$ by use of Lévy statistics, provided that one knows the asymptotic behaviour of the probability distribution for the large values of $\hat{\tau}$. In the case of a one-dimensional cooling scheme, we have obtained the asymptotic distribution of the recycling times as

$$\hat{P}(\hat{\tau}) \underset{\hat{\tau} \gg \hat{\tau}_b}{\simeq} \frac{\hat{\mu}\hat{\tau}_b^{\hat{\mu}}}{\hat{\tau}^{1+\hat{\mu}}} \tag{3.60}$$

in a form similar to eq. (3.32) for the distribution of the trapping times. We have found that

$$\hat{\mu} = \tfrac{1}{2} \tag{3.61}$$

for the case of a homogeneous random walk (*unconfined model*, with a constant delay between successive steps), and

$$\hat{\mu} = \tfrac{1}{4} \tag{3.62}$$

for the case of a jump rate decreasing as p^{-2} (*Doppler model*).

In higher dimensions, $D > 1$, the distribution of return times of an unconfined random walk to the origin becomes extremely large, and the corresponding cooling mechanism is very inefficient. The role of confining walls then becomes crucial.

4

Broad distributions and Lévy statistics: a brief overview

In this chapter, we introduce the main concepts and tools of Lévy statistics that will be used in subsequent chapters in the context of laser cooling. In Section 4.1, we show how statistical distributions with slowly decaying power-law tails can appear in a physical problem. Then, in Section 4.2, we introduce the generalized Central Limit Theorem enabling one to handle statistically 'Lévy sums', i.e. sums of independent random variables, the distributions of which have power-law tails. We also sketch, in a part that can be skipped at first reading, the proof of the theorem and present a few mathematical properties concerning distributions with power-law tails and Lévy distributions. In Section 4.3, we present some properties of Lévy sums which will turn out to be crucial for the physical discussion presented in subsequent chapters: the scaling behaviour, the hierarchy and fluctuation problems. These properties are illustrated using numerical simulations. Finally, in Section 4.4, we present the distribution $S(t)$, called the 'sprinkling distribution'. This distribution presents unexpected features and will play an essential role in the following chapters.

4.1 Power-law distributions. When do they occur?

Situations where broad distributions appear and where rare events play a dominant role are more and more frequently encountered in physics, as well as in many other fields, such as geology, economy and finance. The term 'broad distributions' usually refers to distributions decaying very slowly for large deviations, typically as a power law, implying that some moments of the distribution are formally infinite.

The paradigm problem concerning these types of random variables is the behaviour of the sum of a large number of them. For example, in the problem of interest here, the total experimental time can be decomposed into a sum of the time intervals corresponding to the trapping region and to the external region. Precise theorems govern the properties of these sums, generalizing the well known

(Gaussian) Central Limit Theorem (CLT). We shall not state these results in full generality (the reader can consult [GnK54, BoG90]), but rather focus on the case relevant to our purpose. We shall thus restrict our discussion to *positive* random variables τ (representing random *times*), distributed for large τ as:

$$P(\tau) \underset{\tau \to \infty}{\simeq} \frac{\mu \tau_b^{\mu}}{\tau^{1+\mu}} \tag{4.1}$$

where τ_b sets the scale of the phenomenon, and μ is an exponent describing how fast the distribution decays to 0. (The extra factor μ in the numerator is included for later convenience.) To normalize the distribution P, $\mu > 0$ is required. All the moments $\langle \tau^q \rangle = \int_0^{\infty} d\tau \, \tau^q P(\tau)$ such that $q \geq \mu$ are divergent. The most interesting case, as we shall see below, is the case where $\mu \leq 1$, for which the mean value $\langle \tau \rangle$ of τ is infinite.

When do such power-law distributions occur? They sometimes result from the highly complex underlying dynamics of the physical system, as in chaotic systems [KSZ96, Zas99], and models of avalanches or earthquakes [Bak96, BoC97].

Another frequent scenario for creating power-law distributions is a change of variable. A first variable a, which is naturally sampled by the physical process, is distributed according to a law which may be of any type (Gaussian, exponential, uniform, ...), but the distribution of a related physical quantity $b = f(a)$ turns out to be a power law for certain types of (non-linear) functions $f(a)$. A first example of such a situation was given in Section 3.3. While the probability of reaching a small momentum p is approximately uniform, the lifetime $\tau \propto p^{-\alpha}$ of the corresponding p states is distributed according to a power law, thus leading to eq. (4.1) with $\mu = D/\alpha$.

Another interesting example arising from a change of variable is thermal activation out of a deep potential valley [Shl88, BoD95]. The Arrhenius law states that the average exit time τ is proportional to $\tau_0 \exp(E/k_B T)$, where E is the energy barrier, T the temperature and τ_0 a typical time. In disordered systems, the barriers E are themselves random variables which are often distributed according to an exponential law: $\Pi(E) = E_0^{-1} \exp(-E/E_0)$. The resulting distribution $P(\tau)$ of exit times τ, which is given by $P(\tau)d\tau = \Pi(E)dE$ with $\tau = \tau_0 \exp(E/k_B T)$, is thus equal to

$$
\begin{aligned}
P(\tau) &= \frac{1}{E_0} \exp\left(-\frac{E}{E_0}\right) \frac{k_B T}{\tau_0} \exp\left(-\frac{E}{k_B T}\right) \\
&= \frac{k_B T}{E_0} \frac{\tau_0^{\mu}}{\tau^{1+\mu}}.
\end{aligned} \tag{4.2}
$$

We get an expression similar to (4.1) with $\mu = k_B T / E_0$ and $\tau_b = \tau_0$. Interestingly, for $k_B T < E_0$, the average relaxation time is infinite, leading to strongly anomalous dynamics (see below, and [Bou92, BoD95, BCK97, Bou00]).

Notice that the above derivation of eq. (4.2) assumes that the exit times τ are deterministically fixed by the height E of the barrier. In parallel with the results of Chapter 3 (Section 3.3.1), the result (4.2) is not dramatically altered if the exit times are distributed as an exponential with an average given by the Arrhenius law.

4.2 Generalized Central Limit Theorem

4.2.1 Lévy sums. Asymptotic behaviour and Lévy distributions

Let T_N be the sum of N independent positive random variables, all distributed according to the distribution $P(\tau)$ of eq. (4.1):

$$T_N = \sum_{i=1}^{N} \tau_i. \tag{4.3}$$

When $\mu > 2$, the usual form of the CLT is valid since both the mean value $\langle \tau \rangle$ and the variance $\sigma^2 = \langle \tau^2 \rangle - \langle \tau \rangle^2$ exist. Defining a new variable ξ by

$$T_N = \langle \tau \rangle N + \sigma \sqrt{N}\, \xi, \tag{4.4}$$

the CLT then says that, for large N, ξ tends to a dimensionless Gaussian random variable with zero mean value and unit variance, i.e. it is distributed according to $G(\xi) = (2\pi)^{-1/2} \exp(-\xi^2/2)$ ('normal' distribution). More precisely, one has, independently of the shape of $P(\tau)$:

$$\lim_{N \to \infty} \mathcal{P}\left(\xi_1 \leq \frac{T_N - \langle \tau \rangle N}{\sigma \sqrt{N}} \leq \xi_2 \right) = \int_{\xi_1}^{\xi_2} d\xi\, G(\xi). \tag{4.5}$$

We note that the second (fluctuating) term in eq. (4.4) is negligible compared to the first one when $N \to \infty$.

For $\mu < 2$, the mean value $\langle \tau \rangle$ and/or the variance σ^2 diverge and eq. (4.5) is no longer valid. The CLT has been generalized by Lévy and Gnedenko, and gives results which are *independent of the detailed shape* of $P(\tau)$ and which depend only on the long time behaviour described by eq. (4.1). The sums T_N are called 'Lévy sums'. We now state a few important results concerning the asymptotic behaviour of these Lévy sums (for large N). A sketch of the proof of these results will be presented in the next section, using the properties of the Laplace transforms of functions with power-law tails.

The generalized CLT takes two different forms for $1 < \mu < 2$ and for $\mu < 1$[1].

---

[1] Logarithmic corrections appear in the cases $\mu = 1$ and $\mu = 2$, requiring a separate discussion (see Appendix C).

Consider first the case $1 < \mu < 2$, where τ has a finite mean value $\langle \tau \rangle$ but an infinite variance. If we introduce a new variable ξ by

$$1 < \mu < 2: \qquad T_N = \langle \tau \rangle N + \xi \tau_b N^{1/\mu}, \tag{4.6}$$

then the generalized CLT states that ξ is a random variable of order one, distributed for large N according to a function $L_\mu(\xi)$ which depends only on μ and which is called the 'completely asymmetric' Lévy distribution of index μ[2]. More precisely, we can write

$$\lim_{N \to \infty} \mathcal{P}\left(\xi_1 \le \frac{T_N - \langle \tau \rangle N}{\tau_b N^{1/\mu}} \le \xi_2 \right) = \int_{\xi_1}^{\xi_2} \mathrm{d}\xi\, L_\mu(\xi). \tag{4.7}$$

Note that the second (fluctuating) term in eq. (4.6) is still negligible compared with the first when $N \to \infty$. The Lévy distributions $L_\mu(\xi)$ have simple Laplace transforms[3]:

$$\mathcal{L}L_\mu(u) = \int_0^\infty \mathrm{d}\xi\, L_\mu(\xi) \mathrm{e}^{-u\xi} = \exp(-b_\mu u^\mu) \quad \text{with} \quad b_\mu = \frac{(\mu - 1)\Gamma(1 - \mu)}{\mu}. \tag{4.8}$$

In the case $\mu < 1$, both the mean value and the variance of τ diverge and one finds that T_N grows faster than the number of terms N. Equation (4.6) has to be replaced by

$$\mu < 1: \qquad T_N = \xi \tau_b N^{1/\mu}, \tag{4.9}$$

and one finds that ξ is again a random variable of order one, distributed for large N according to a Lévy distribution $L_\mu(\xi)$, whose Laplace transform is now:

$$\mathcal{L}L_\mu(u) = \int_0^\infty \mathrm{d}\xi\, L_\mu(\xi) \mathrm{e}^{-u\xi} = \exp(-b_\mu u^\mu) \quad \text{with} \quad b_\mu = \Gamma(1 - \mu). \tag{4.10}$$

The analogue of eq. (4.7) is:

$$\lim_{N \to \infty} \mathcal{P}\left(\xi_1 \le \frac{T_N}{\tau_b N^{1/\mu}} \le \xi_2 \right) = \int_{\xi_1}^{\xi_2} \mathrm{d}\xi\, L_\mu(\xi). \tag{4.11}$$

4.2.2 *Sketch of the proof of the generalized CLT*

We try now to give an idea of the mathematical properties leading to the very simple forms (4.8) and (4.10) for the Laplace transforms of the Lévy distributions $L_\mu(\xi)$.

The fact that Laplace transforms play an important role in this problem is easy

---

[2] Since τ is positive, L_μ is actually a particular case ('completely asymmetric') of more general Lévy distributions, which arise when the random variable involved in the summation has power-law tails both at $+\infty$ and at $-\infty$.

[3] We denote the Laplace transform of f by $\mathcal{L}f$.

to understand. Let $\Pi_N(T_N)$ be the probability distribution of the Lévy sum T_N. It can be written:

$$\Pi_N(T_N) = \int d\tau_1 \ldots d\tau_N \, P(\tau_1) \ldots P(\tau_N) \, \delta \left(\sum_{i=1}^{N} \tau_i - T_N \right) \tag{4.12}$$

where the constraint on the value of the sum is imposed through a δ-function. In fact, the right-hand side of eq. (4.12) is a convolution product of N functions $P(\tau)$, so that the Laplace transform $\mathcal{L}\Pi_N(s)$ of $\Pi_N(T_N)$ is nothing but the Nth power of the Laplace transform $\mathcal{L}P(s)$ of $P(\tau)$[4]:

$$\mathcal{L}\Pi_N(s) = \left[\int_0^\infty d\tau \, P(\tau) e^{-s\tau} \right]^N = [\mathcal{L}P(s)]^N. \tag{4.13}$$

We now use the fact that $P(\tau)$ is a probability distribution, i.e. takes positive values and is normalized to one. This implies that $\mathcal{L}P(s) \leq 1$ for any $s \geq 0$, the upper bound being obtained for $s = 0$. Since $\mathcal{L}P(s)$ is raised to a high power N in eq. (4.13), one expects that $\mathcal{L}\Pi_N(s)$, which is equal to one for $s = 0$, will be appreciable only in the neighbourhood of $s = 0$. This explains the importance in this problem of the small-s behaviour of $\mathcal{L}P(s)$, which is itself determined by the long-τ behaviour of $P(\tau)$.

We will focus here on distributions (4.1) with $\mu < 1$. We suppose in addition that the subleading corrections to eq. (4.1) decay faster than τ^{-2} for large τ. One can then show that the small-s behaviour of their Laplace transforms $\mathcal{L}P(s)$ is given by

$$\mathcal{L}P(s) \underset{s\to 0}{=} 1 - \Gamma(1-\mu)(\tau_b s)^\mu - A_0 \tau_b s + \cdots \tag{4.14}$$

where A_0 is a constant. In view of its importance here, a brief proof of this result will be given in point (ii) of Section 4.2.3.

Using eq. (4.13), one gets:

$$\mathcal{L}\Pi_N(s) \underset{s\to 0}{=} \left[1 - \Gamma(1-\mu)(\tau_b s)^\mu + O(\tau_b s) \right]^N. \tag{4.15}$$

Setting $\hat{s} = s\,\tau_b N^{1/\mu}$, one obtains

$$\mathcal{L}\Pi_N \left(s = \frac{\hat{s}}{\tau_b N^{1/\mu}} \right) \underset{s\to 0}{=} \left(1 - \Gamma(1-\mu)\frac{\hat{s}^\mu}{N} + \frac{O(\hat{s})}{N^{1/\mu}} \right)^N. \tag{4.16}$$

---

[4] Note that in eq. (4.13) s is conjugate to a time variable, T_N or τ, so that it has the dimension of the inverse of time, whereas in eqs. (4.8) and (4.10) the conjugate variables ξ and u are both dimensionless.

Taking the limit $N \to \infty$ and $s \to 0$, with $\hat{s}$ fixed, gives

$$\mathcal{L}\Pi_N\left(s = \frac{\hat{s}}{\tau_b N^{1/\mu}}\right) \underset{s \to 0}{=} \exp\left[N \ln\left(1 - \Gamma(1-\mu)\frac{\hat{s}^\mu}{N} + \frac{O(\hat{s})}{N^{1/\mu}}\right)\right]$$

$$\underset{N \to \infty}{\longrightarrow} \exp\left(-\Gamma(1-\mu)\hat{s}^\mu + \frac{O(\hat{s})}{N^{1/\mu - 1}}\right)$$

$$\underset{N \to \infty}{\longrightarrow} \exp\left[-\Gamma(1-\mu)\hat{s}^\mu\right], \tag{4.17}$$

since $\mu < 1$. Using the definition of $\mathcal{L}\Pi_N(s) = \int e^{-sT_N} \Pi_N(T_N)\,dT_N$, the change of variable $\xi = T_N/\tau_b N^{1/\mu}$ and the relation $P(\xi)\,d\xi = \Pi_N(T_N)\,dT_N$, the above calculation directly shows that $\mathcal{L}L_\mu(u)$ given by eq. (4.10) is indeed the Laplace transform of the distribution of ξ at large N.

4.2.3 A few mathematical results

We gather in this subsection a few useful mathematical results which are referred to in this chapter. This part can be skipped at first reading.

A few properties of the Laplace transforms of functions with power-law tails

(i) Suppose first that $\mu > 1$ so that $\langle \tau \rangle$ is finite. For $s \to 0$, one can then write

$$\mathcal{L}P(s) = \int_0^\infty d\tau\, P(\tau)e^{-s\tau}$$

$$\underset{s \to 0}{\simeq} \int_0^\infty d\tau\, P(\tau)(1 - s\tau) = 1 - s\langle\tau\rangle. \tag{4.18}$$

We will come back to the higher-order terms of the small-s expansion of $\mathcal{L}P(s)$ (see eq. (4.23)).

(ii) If $\mu < 1$, the previous expression is no longer valid because $\langle \tau \rangle$ is infinite. We rewrite $e^{-s\tau}$ in the first line of eq. (4.18) as $1 + e^{-s\tau} - 1$, so that

$$\mathcal{L}P(s) = \int_0^\infty d\tau\, P(\tau)(1 + e^{-s\tau} - 1)$$

$$= 1 + \int_0^\infty d\tau\, P(\tau)(e^{-s\tau} - 1). \tag{4.19}$$

Let τ^* be the value of τ beyond which the asymptotic expression (4.1) is correct. The integral of the last line of eq. (4.19) from 0 to ∞ can be split into an integral from 0 to τ^* and an integral from τ^* to ∞. Since $\left|e^{-s\tau} - 1\right| < s\tau$, one has:

$$\left|\int_0^{\tau^*} d\tau\, P(\tau)(e^{-s\tau} - 1)\right| < s\int_0^{\tau^*} d\tau\, \tau\, P(\tau) < \tau^* s \int_0^{\tau^*} d\tau\, P(\tau)$$

$$< \tau^* s \int_0^\infty d\tau\, P(\tau) < \tau^* s. \tag{4.20}$$

Thus, when s tends to 0, more precisely when $s \ll 1/\tau^*$, the integral from 0 to τ^* is at most of order $O(\tau^* s)$. In the integral from τ^* to ∞, we replace $P(\tau)$ by its asymptotic form (4.1) and we perform integration by parts. This gives, putting $x = s\tau$:

$$\int_{\tau^*}^{\infty} d\tau \, P(\tau)(e^{-s\tau} - 1) = \mu(\tau_b s)^{\mu} \int_{\tau^* s}^{\infty} dx \, x^{-(1+\mu)}(e^{-x} - 1)$$

$$= (\tau_b s)^{\mu}(e^{-\tau^* s} - 1)(\tau^* s)^{-\mu} - (\tau_b s)^{\mu} \int_{s\tau^*}^{\infty} dx \, x^{-\mu} e^{-x}. \qquad (4.21)$$

Combining the last line of eq. (4.21) with eq. (4.20), we obtain

$$\mathcal{L}P(s) \underset{s \to 0}{\simeq} 1 - \Gamma(1 - \mu)(\tau_b s)^{\mu} - A_0 \tau_b s + \cdots \qquad (4.22)$$

where A_0 is a constant depending on the detailed shape of $P(\tau)$. This is nothing but eq. (4.14).

If we subtract from $P(\tau)$ its asymptotic behaviour (4.1), we are left with a new function $\widetilde{P}(\tau)$ which decays faster than $\tau^{-(1+\mu)}$ at large τ. If it decays faster than τ^{-2}, the integral $\int_0^{\infty} d\tau \, \tau \, \widetilde{P}(\tau)$ converges and a calculation similar to that of eq. (4.18) gives a term of order $O(s\,\tau^*)$ when $s \to 0$. Combined with similar contributions of the same order from eq. (4.20) and eq. (4.21), this gives the last term of the right-hand side of eq. (4.22).

(iii) If μ had been larger than one, but different from any integer[5], the small-s expansion of $\mathcal{L}P(s)$ would have taken the following form:

$$\mathcal{L}P(s) = 1 - M_1 s + \frac{M_2}{2!} s^2 + \cdots + (-1)^n \frac{M_n}{n!} s^n - C_\mu s^\mu - \cdots \qquad (4.23)$$

where n is the integer value of μ, and the M_i are the moments of $P(\tau)$ (for example, $M_1 = \langle \tau \rangle$ is the mean value of τ). In other words, the small-s expansion is regular up to its nth term, until the power-law singularity is met. Conversely, the knowledge of $\mathcal{L}P(s)$ for small s allows one to extract the power-law behaviour of $P(\tau)$ for large τ.

(iv) Actually, one should also note that eq. (4.14) can be extended to the case where $P(\tau)$ is not a normalizable probability density and varies as $C\tau^{-(1-\nu)}$ at large τ with $\nu > 0$. Such a case was encountered in Chapter 3, Section 3.4.2: the probability of an atom being present at the starting point of a three-dimensional random walk decays as $\tau^{-1/2}$, corresponding to $\nu = 1/2$. In this case, calculations similar to the previous ones show that the leading term of $\mathcal{L}P(s)$ for small s reads:

$$\mathcal{L}P(s) = \Gamma(\nu)Cs^{-\nu} + A + \cdots \qquad (4.24)$$

where A is a constant, again depending on the detailed shape of $P(\tau)$.

A few properties of Lévy distributions

We now list without proofs a few important properties of $L_\mu(\xi)$, defined in eqs. (4.8) and (4.10), remembering that we are restricting ourselves to the case of positive random variables.

---

[5] Again, if μ is an integer, logarithmic corrections appear, see Appendix C.

(i) For $\mu = 2$, $L_\mu(\xi)$ reduces to the usual Gaussian distribution $G(\xi) = (2\pi)^{-1/2} \exp(-\xi^2/2)$.

(ii) For $0 < \mu < 2$ and $\xi \to \infty$, $L_\mu(\xi)$ decays as a power law with the same exponent as $P(\tau)$:

$$L_\mu(\xi) \underset{\xi \to +\infty}{\simeq} \frac{\mu}{\xi^{1+\mu}} + O\left(\frac{1}{\xi^{1+2\mu}}\right). \tag{4.25}$$

(iii) For $\mu < 1$, $L_\mu(\xi)$ is obviously 0 for $\xi < 0$ and has an essential singularity for $\xi \to 0$:

$$L_\mu(\xi) \underset{\xi \to 0}{\simeq} A\, \xi^{\frac{\mu-2}{2(1-\mu)}} \exp\left(-B\, \xi^{\frac{\mu}{\mu-1}}\right) \tag{4.26}$$

where A and B are prefactors.

(iv) For $\mu = 1/2$, an explicit expression can be given for all ξ:

$$L_{1/2}(\xi) = Y(\xi) \frac{1}{2\xi^{3/2}} \exp\left(-\frac{\pi}{4\xi}\right) \tag{4.27}$$

where $Y(\xi)$ is the Heaviside function. The variations of $L_{1/2}(\xi)$ with ξ are represented in figure 4.1. All functions $L_\mu(\xi)$ with $\mu < 1$ have qualitatively similar variations. Note that the maximum of $L_{1/2}(\xi)$ is reached for $\xi = \pi/6$, which clearly shows that the dimensionless random variable ξ is of the order of one.

(v) For $1 < \mu < 2$, $L_\mu(\xi)$ describes the fluctuations of T_N around the mean value $N\langle\tau\rangle$, and thus extends from $-\infty$ to $+\infty$. The decay of $L_\mu(\xi)$ for $\xi \to -\infty$ is however much faster than the power law (4.25), and is given by:

$$L_\mu(\xi) \underset{\xi \to -\infty}{\simeq} C\, \xi^{\frac{\mu-2}{2(1-\mu)}} \exp\left(-D\, |\xi|^{\frac{\mu}{\mu-1}}\right) \tag{4.28}$$

where C and D are prefactors.

(vi) Only the moments of order $q < \mu$ of $L_\mu(\xi)$ exist. For $\mu < 1$, an explicit calculation leads to:

$$\langle \xi^q \rangle \equiv \int_0^\infty d\xi\, \xi^q L_\mu(\xi) = b_\mu^{q/\mu} \frac{\Gamma(-q/\mu)}{\mu\Gamma(-q)} \tag{4.29}$$

where $b_\mu = \Gamma(1-\mu)$ (see eq. (4.10)).

4.3 Qualitative discussion of some properties of Lévy sums

4.3.1 Dependence of a Lévy sum on the number of terms for $\mu < 1$

One of the most important results of the generalized CLT is that a Lévy sum T_N scales as $N^{1/\mu}$ when $\mu < 1$ (see eq. (4.9)). For example, for $\mu = 1/2$, T_N scales as N^2; for $\mu = 1/4$, as N^4. The smaller μ, the greater the exponent of the power-law dependence of T_N on N. Such behaviour is quite different from that of usual random variables τ which have a finite mean value $\langle\tau\rangle$ and for which T_N scales as $N\langle\tau\rangle$ (usual law of large numbers).

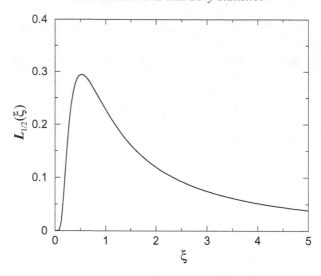

Fig. 4.1. Function $L_{1/2}(\xi)$. After a very slow increase near $\xi = 0$, $L_{1/2}(\xi)$ reaches a maximum for $\xi = \pi/6$ and then decreases as a power law at large ξ (as $1/(2\xi^{3/2})$).

Such a result is linked to the fact that the probability of having a very large value of τ in a drawing of the random variable is not negligible when $P(\tau)$ decreases slowly at large τ. When one increases the number N of trials, larger and larger values of τ can be obtained, and this explains why the sum T_N can grow faster than N.

4.3.2 Hierarchical structure in a Lévy sum

We now want to address the following questions. Suppose that one orders the sequence $\{\tau_1, \tau_2, \ldots, \tau_N\}$ of the various terms of a Lévy sum T_N from the largest one to the smallest. Let $\tau^{(1)}$ be the first one (the largest), $\tau^{(2)}$ the next one, $\ldots$, $\tau^{(n)}$ the nth one. What are the orders of magnitude of these various terms? How do they scale with N and with n? In other words, is there a hierarchy between these terms? Is $\tau^{(n)}$ much larger than $\tau^{(n+1)}$?

To answer these questions, we first determine the most probable value of $\tau^{(n)}$. Let $\Pi(\tau^{(n)})d\tau$ be the probability of finding the nth term between $\tau^{(n)}$ and $\tau^{(n)} + d\tau$. We have (see also Section 2.1.1 in [Gum58]):

$$\Pi(\tau^{(n)}) = N \binom{N-1}{n-1} P(\tau^{(n)}) \left[\int_{\tau^{(n)}}^{\infty} d\tau\, P(\tau) \right]^{n-1} \left[1 - \int_{\tau^{(n)}}^{\infty} d\tau\, P(\tau) \right]^{N-n}.$$

(4.30)

The first term, N, corresponds to the N possible positions of $\tau^{(n)}$ in the sequence

$\tau_1, \tau_2, \ldots, \tau_N$. The second term, $\binom{N-1}{n-1}$, counts the different possible ways of obtaining $n - 1$ drawings larger than $\tau^{(n)}$ and $N - n$ smaller than $\tau^{(n)}$. Finally, the last three terms are the probabilities of drawing values of τ equal to, larger or smaller than $\tau^{(n)}$, respectively, raised to the appropriate power. Using eq. (4.1), one gets

$$\int_{\tau^{(n)}}^{\infty} d\tau \, P(\tau) = \left(\frac{\tau_b}{\tau^{(n)}} \right)^{\mu} \tag{4.31}$$

and a simple calculation shows that the most probable value of $\tau^{(n)}$, which maximizes eq. (4.30), is given by:

$$\tau^{(n)} = \tau_b \left[\frac{1 + \mu N}{1 + \mu n} \right]^{1/\mu}$$
$$\simeq \tau_b \left(\frac{N}{n} \right)^{1/\mu} \quad \text{if} \quad N, n \gg 1/\mu. \tag{4.32}$$

A first important result expressed by eq. (4.32) is that the largest term of a Lévy sum, $\tau^{(1)}$, scales with N as $\tau_b N^{1/\mu}$. This result is valid for any value of $\mu > 0$, in the limit $N \to \infty$. Interestingly, for $\mu < 1$, one has $T_N \simeq \tau_b N^{1/\mu}$ according to eq. (4.9) so that the largest term $\tau^{(1)}$ is of the order of the sum itself. A single term of the Lévy sum can be of the order of the total sum! This is the most important qualitative property of the Lévy sums for $\mu < 1$: a significant fraction of the total 'time' T_N is spent in the 'deepest trap'. This is precisely the situation encountered in the Monte Carlo simulations described in Chapter 2.

The n-dependence of $\tau^{(n)}$ is also very interesting. As soon as n becomes larger than $1/\mu$, $\tau^{(n)}$ scales with n as $n^{-1/\mu}$. For example, for $\mu = 1/2$, $\tau^{(10)}$ is $2^2 = 4$ times larger than $\tau^{(20)}$, $3^2 = 9$ times larger than $\tau^{(30)}$, and so on. In other words, there is a strong hierarchy between the various terms of a Lévy sum with $\mu < 1$. Such a sum is 'dominated' by a very small number of terms. If one plots $\ln \tau^{(n)}$ versus $\ln n$, one expects, according to eq. (4.32), to get a straight line with a slope $-1/\mu$. Conversely, when one analyses a set of independent random numbers, it may be useful to order them and to plot $\ln \tau^{(n)}$ versus $\ln n$. If one gets a straight line with a slope $-1/\mu$, this is a good indication that the corresponding random variable is distributed according to a power-law distribution such as eq. (4.1)[6].

It is interesting to compare the previous results, typical of Lévy statistics, with those corresponding to usual Gaussian statistics where $P(\tau)$ is a 'narrow' distribution for which the CLT is applicable. Take, for example, the exponential

---

[6] A more precise 'maximum likelihood' procedure to estimate the exponent μ is known as the Hill estimator, see [Hil75].

distribution

$$P(\tau) = \frac{1}{\tau_b}\, e^{-\tau/\tau_b} \tag{4.33}$$

leading to a simple analytical expression for the value of $\tau^{(n)}$ which maximizes eq. (4.30):

$$\tau^{(n)} = \tau_b \ln\left(\frac{N}{n}\right). \tag{4.34}$$

Instead of power-law variations with N and n, we obtain now logarithmic variations which are extremely slow. In other words, there is now no hierarchy between the various terms of the sum which are all of the same order. An increase of the size N of the statistical sample leads only to a modest increase of the typical size of the largest term $\tau^{(1)}$.

4.3.3 Large fluctuations

For usual statistics obeying the standard CLT (finite $\langle \tau \rangle$ and $\langle \tau^2 \rangle$), the sample to sample fluctuations of the sum T_N vanish when the size of the sample, i.e. the number of terms N, increases. More precisely, let us consider the relative fluctuations $\sigma_r(N)$ of the average value for a sample of size N defined by[7]:

$$\sigma_r(N) = \frac{\langle |(T_N/N) - \langle \tau \rangle| \rangle}{\langle \tau \rangle} = \frac{\langle |T_N - N\langle \tau \rangle| \rangle}{N\langle \tau \rangle}. \tag{4.35}$$

According to eqs. (4.4) and (4.5), the variable $(T_N - N\langle \tau \rangle)/(\sigma\sqrt{N})$ is of order one when $N \gg 1$ and a simple calculation leads to

$$\langle \tau^2 \rangle < \infty: \quad \sigma_r(N) \simeq \frac{\sigma}{\langle \tau \rangle \sqrt{N}}. \tag{4.36}$$

These fluctuations tend to zero when N tends to infinity. This guarantees an asymptotically perfect repeatability of average values in the limit of large samples. In other words, average values can be accurately predicted for large samples, even if individual values fluctuate a lot. This is the origin of the traditional success of statistical methods in both natural and social sciences.

For Lévy statistics, the situation can be radically different. For $1 < \mu < 2$, a simple calculation using eq. (4.6) leads to

$$1 < \mu < 2: \quad \sigma_r(N) \simeq \tau_b/(\langle \tau \rangle N^{1-1/\mu}). \tag{4.37}$$

The relative fluctuations of the average value again vanish[8] at large N, although

---

[7] We take an absolute value instead of a root mean square to avoid divergencies for the case $1 < \mu < 2$ considered below.

[8] But the fluctuations of the second moment would not vanish.

more slowly than $1/\sqrt{N}$. But for $0 < \mu < 1$, this is no longer true. Since we can no longer define the relative fluctuations $\sigma_r(N)$ by (4.35) ($\langle\tau\rangle$ is infinite), we use the following argument: as the largest term $\tau^{(1)}$ is of the order of the sum T_N, the sum T_N fluctuates as much as a single term. Therefore the relative fluctuations from sample to sample are the same as the fluctuations from term to term, i.e. they are of order one whatever the size of the sample:

$$\mu < 1: \quad \sigma_r(N) \simeq 1. \tag{4.38}$$

As a consequence, the value of the sum T_N is not repeatable from one sample to another sample. The accuracy of the statistical prediction is not improved by increasing the sample size.

It thus appears that Lévy statistics lead, when $\mu < 1$, to a behaviour which is radically different from that deduced from the usual CLT [Man82, Man96]. The usual CLT describes how the fluctuations vanish at large N, whereas the generalized CLT (for $\mu < 1$) shows that the fluctuations continue to play an essential role however large N may be.

Repeatability is unavoidably lost when $\mu < 1$. However, the generalized CLT still allows *some predictability*. It predicts the typical, i.e. most probable, values for the sums T_N. Such an order of magnitude prediction is the best that statistical tools can offer when $\mu < 1$.

It is worth pointing out that the presence in a physical phenomenon of a sum T_N undergoing large fluctuations does not necessarily imply that the phenomenon is on the whole unrepeatable. Other quantities related to, but different from, T_N can still be accurately predicted even when $\mu < 1$. The physically relevant quantities calculated in the following chapters are of this kind.

4.3.4 Illustration with numerical simulations

All the spectacular features of Lévy statistics analysed in the previous section clearly appear in numerical simulations. These numerical simulations are performed in the following way. One makes successive drawings $\tau_1, \tau_2, \ldots, \tau_N, \ldots$ of the random variable τ distributed according to eq. (4.1), and one plots $T_N = \sum_{i=1}^{N} \tau_i$ versus N, for different values of μ. These sequences are generated using the *same*[9] sequence $x_1, x_2, \ldots, x_N, \ldots$ of random numbers uniformly distributed between 0 and 1, and then defining:

$$\tau_i = \tau_b x_i^{-1/\mu}. \tag{4.39}$$

---

[9] The use of the same sequence of x_i enables one to see the effects of different μ values not blurred by the statistical fluctuations.

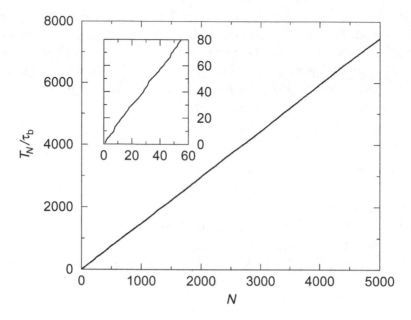

Fig. 4.2. Plot of T_N (in units of τ_b) versus N for $\mu = 3$. The inset shows a zoom of a small portion of the plot.

One can check that this transformation produces τ_i values that are distributed according to eq. (4.1).

Figure 4.2 shows T_N (in units of τ_b) versus N for $\mu = 3$. In this case, $\langle \tau \rangle$ is finite and equal to $\mu \tau_b/(\mu - 1) = 1.5\,\tau_b$ (see eq. (3.35)), and one obtains a plot which looks like a straight line with a slope $\mu/(\mu - 1) = 1.5$. In fact, there are $N = 5000$ vertical steps in such a plot, but each individual step is so small that it cannot be distinguished in the full scale figure. Zooming in on a small portion of the figure reveals these individual steps which appear to be all of the same order (see inset of figure 4.2).

For $\mu < 1$, when $\langle \tau \rangle$ is infinite, the plot has a radically different shape. It looks like a 'devil's staircase' where a small number of individual large steps are clearly visible and are of the order of the total sum itself (see, for example, figure 4.3 corresponding to $\mu = 1/2$). When μ is still smaller, for example when $\mu = 0.1$, one nearly sees only a single huge step (see figure 4.4). Between two large steps, T_N remains nearly constant. This is due to the strong hierarchy between the individual steps (see eq. (4.32)). A few of them are so large that the others can hardly be distinguished. Note the difference of the vertical scales from figure 4.2 to figure 4.3 and figure 4.4, which reflects the $N^{1/\mu}$ dependence of T_N when $\mu < 1$. Zooming in on a small portion of figure 4.3 and figure 4.4 reveals a structure which

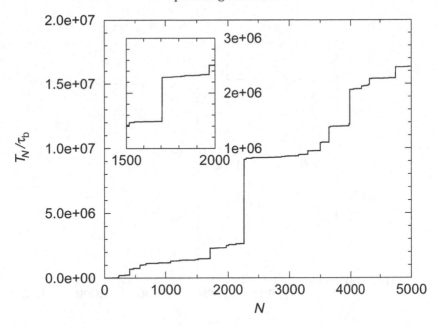

Fig. 4.3. As figure 4.2, but with $\mu = 1/2$. Note the difference of vertical scale. Contrary to figure 4.2, a few large steps are clearly visible and are of the order of the total sum. The same general behaviour appears in the zoom shown in the inset.

has the same shape as the full scale figures (see the insets): one still gets a kind of 'devil's staircase' dominated by a small number of large steps. In other words, the behaviour of T_N versus N is self-similar at all scales.

The hierarchical structure of the various terms of a Lévy sum also appears in rank ordered histograms where one plots $\ln \tau^{(n)}$ versus $\ln n$. Figure 4.5 shows such plots for $\mu = 3$ and $\mu = 1/2$. As expected from the calculations of Section 4.3.2, one obtains a decrease which is well represented by a straight line with a slope equal to $-1/\mu$. These straight lines are shown as interrupted lines in the figure. Note that for $\mu = 1/2$ there are about six orders of magnitude between the largest term and the smallest term of the sequence.

4.4 Sprinkling distribution

4.4.1 Definition. Laplace transform

In this section, we introduce a probability distribution which will be useful for the calculations presented in Chapters 5 and 6. Suppose that one makes successive drawings $\tau_1, \tau_2, \ldots, \tau_n, \ldots$ of the random variable τ distributed according to eq. (4.1), and let us define a random sequence of events $M_1, M_2, \ldots, M_n, \ldots$

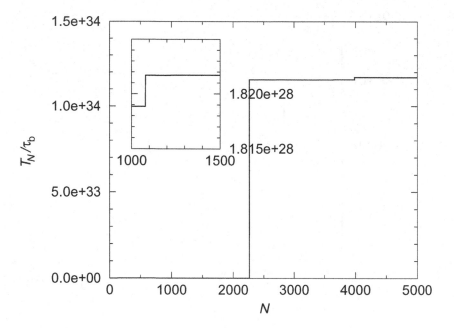

Fig. 4.4. As figure 4.2, but with $\mu = 0.1$. The hierarchical structure in the Lévy sum is still more pronounced than in figure 4.3 and the largest term is huge (note the new vertical scale) and dominates all the others. Here also, the same behaviour appears at all scales (see the inset).

occurring at times $t_1, t_2, \ldots, t_n, \ldots$ such that

$$t_1 = \tau_1, \quad t_2 = \tau_1 + \tau_2, \ldots, \quad t_n = t_{n-1} + \tau_n, \ldots. \tag{4.40}$$

In other words, we introduce a random set of events such that the time intervals between two *successive* events is distributed according to $P(\tau)$. This is illustrated in Fig. 4.6. Averaging over several different realizations of such a random sequence, one can then ask the following question: what is the probability density $S(t)$ of finding an event at time t, disregarding the number of previous events? We shall call such a distribution the 'sprinkling distribution' associated with $P(t)$. It represents the mean density at time t of the random sequence of events $M_1, M_2, \ldots, M_n, \ldots$ introduced above.

It is easy to find an equation satisfied by $S(t)$. Either the event observed at time t is the first one which appears, with probability $P(t)$; or an arbitrary number of events have already occurred before this event, the last one happening at $t_l < t$. Hence, one has:

$$S(t) = P(t) + \int_0^t dt_l\, P(t - t_l)S(t_l). \tag{4.41}$$

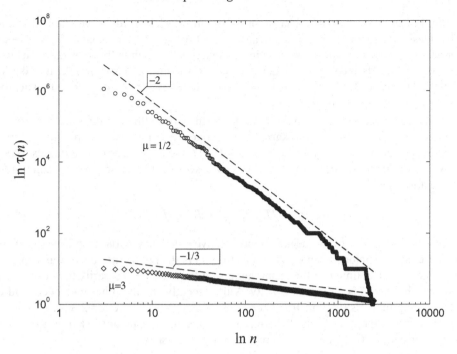

Fig. 4.5. Rank ordered histograms giving $\ln \tau^{(n)}$ versus $\ln n$ for two different values of μ: $\mu = 3$ and $\mu = 1/2$. The interrupted straight lines give the theoretically predicted behaviour of a linear decrease with a slope equal to $-1/\mu$.

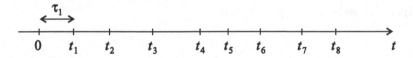

Fig. 4.6. Random set of events M_i occurring at times t_i, with a distribution $P(\tau)$ for the time intervals between two successive events.

This equation is readily solved using Laplace transforms, which converts the convolution into a simple product and one gets:

$$\mathcal{L}S(s) = \frac{\mathcal{L}P(s)}{1 - \mathcal{L}P(s)}. \tag{4.42}$$

4.4.2 Examples taken from other fields

In this book, the events that will be considered in the next sections, and which will be characterized by the sprinkling distribution $S(t)$, are the successive entries of the atom in the trapping zone $p \leq p_{\text{trap}}$ during its random walk in momentum space.

In the general theory of stochastic processes, the sprinkling distribution is known as the density of a renewal process, or the 'renewal density'[10]. Schematically, a renewal process is a statistical process in which a device, say a lightbulb, is installed at time $t = 0$, until it fails and is replaced at time $t = \tau_1$ (random variable), until the new device fails and is replaced at time $t = \tau_1 + \tau_2, \ldots$. The renewal density indicates (statistically) when the devices must be replaced.

Renewal processes are ubiquitous in quantum optics although they are not usually named as such. Consider, for example, the sequence of fluorescence photons emitted by a single atom excited by a resonant laser field (single atom resonance fluorescence). A very important quantity characterizing such a sequence is the so-called second-order correlation function

$$G_2(t) = \langle E^-(0)E^-(t)E^+(t)E^+(0) \rangle \tag{4.43}$$

where E^- and E^+ are the negative and positive frequency parts, respectively, of the electric field operator. It can be shown that $G_2(t)$ is the probability of having a sponta-neous emission at time $t = 0$ and another one at time t, not necessarily the next one. It is clear that $G_2(t)$ is a renewal density. More recently, attention has also been paid to the waiting time distribution $W(\tau)$ (or 'delay function', see Section 2.3.3), giving the distribution of the time intervals between two *successive* spontaneous emissions. The two distributions $G_2(t)$ and $W(\tau)$ are related by an equation

$$G_2(t) = W(t) + \int_0^t dt_l \, W(t - t_l)G_2(t_l), \tag{4.44}$$

which is identical to the renewal equation (4.41) giving the sprinkling distribution, with the correspondence $W \to P$ and $G_2 \to S$ (see, for example, eq. (6.19) in [Rey83], eq. (4.13) in [RDC88] or eq. (45) in [PlK98]).

4.4.3 Asymptotic behaviour. Broad versus narrow distributions

We now investigate the long time behaviour of $S(t)$. Suppose first that $\mu > 1$ so that $\langle \tau \rangle$ is finite. Using eq. (4.18), which states that the small-s expansion of $\mathcal{L}P(s)$ is $1 - \langle \tau \rangle s + \cdots$, one gets:

$$\mathcal{L}S(s) \underset{s \to 0}{\simeq} \frac{1}{\langle \tau \rangle} \frac{1}{s} - 1 \cdots . \tag{4.45}$$

This shows that, for large times, $S(t)$ is constant, equal to $1/\langle \tau \rangle$. We thus find an *a priori* obvious result. For large times, the probability of finding a particular event between t and $t + dt$ is a constant equal to the inverse of the average time interval $\langle \tau \rangle$ between two successive events. In other words, the set of events M_1, $M_2, \ldots, M_n, \ldots$ has a constant density equal to $1/\langle \tau \rangle$.

---

[10] This connection was made in [BaB00], see Section 10.2.1.

Equation (4.45) is in fact valid for any probability distribution $P(\tau)$ with a finite average value. This is in particular the case of the sprinkling distribution $G_2(t)$ associated with single atom resonance fluorescence, since $W(t)$, which plays the role of $P(t)$, is a sum of exponentials with a finite mean value. Another simple example is provided by the Poisson process which enables us to check eq. (4.45). In this case, the waiting time distribution is $P(\tau) = \Gamma e^{-\Gamma\tau}$ and one expects the rate $S(t)$ of occurrence of the events to be constant, for any time including small times. Using eq. (4.42) and $\mathcal{L}P(s) = \Gamma/(\Gamma + s)$, one obtains $\mathcal{L}S(s) = \Gamma/s$. This agrees with the first term of eq. (4.45) but, in this case, the relation is exact, for any s. We get $S(t) = \Gamma$ for all t, as expected for a Poisson process.

Such a result is no longer valid when $\mu < 1$. We must now use eq. (4.14) which, inserted into eq. (4.42), gives[11]

$$\mathcal{L}S(s) \underset{s\to 0}{\simeq} \frac{1}{\Gamma(1-\mu)}(\tau_b s)^{-\mu} + \text{subleading terms.} \tag{4.46}$$

Using eq. (4.24) and the identity

$$\Gamma(\mu)\Gamma(1-\mu) = \frac{\pi}{\sin(\pi\mu)}, \tag{4.47}$$

one finally gets[12]

$$S(t) \underset{t\to\infty}{\simeq} \frac{\sin(\pi\mu)}{\pi}\frac{1}{\tau_b}\left(\frac{\tau_b}{t}\right)^{1-\mu} + O\left[(\tau_b/t)^{2-2\mu}\right]. \tag{4.48}$$

Note that $S(t)$ has the dimension of the inverse of a time, but goes to zero when $t \to \infty$. This is related to the fact that, as the time t increases, the probability of drawing a large value of τ, of the order of t itself, remains constant, so that the mean density of events decreases. In such a process, *the rate of events decreases at long times due to a purely statistical property* ($\langle\tau\rangle = \infty$) *while the distribution $P(\tau)$ of the increments τ_i is perfectly stationary*. The identification of this unusual feature in laser cooling is one of the most salient results of the presented statistical approach.

This also means that the observation of $S(t)$ allows one to infer the starting 'date' ($t = 0$) of the process – which would of course be impossible to do for $\mu > 1$. In other words, time translation invariance is broken for $\mu < 1$ and the process 'ages'. Such a scenario was discussed in the context of glassy dynamics in [Bou92, BoD95, BCK97]. The sprinkling distribution $S(t)$ associated with a broad distribution $P(t)$ therefore exhibits interesting new features compared with the usual case where $P(t)$ has a finite mean value.

---

[11] If $\mu < 1/2$, the subleading terms of eq. (4.46) are constant terms plus terms in $(\tau_b s)^{1-2\mu}$. If $\mu > 1/2$, these corrections are in $(\tau_b s)^{1-2\mu}$.

[12] If $\mu = 1$, logarithmic terms appear.

5

The proportion of atoms trapped in quasi-dark states

We now have all the mathematical tools in hand to address the important questions for the cooling process, namely: what is the proportion $f_{\text{trap}}(\theta)$ of 'trapped' atoms (i.e. those which have a very small momentum $p < p_{\text{trap}}$); what is the 'line shape', i.e. the momentum distribution, after an interaction time θ?

In Section 5.1, we define precisely the trapped proportion $f_{\text{trap}}(\theta)$ in terms of an ensemble average and compare it to a time average defined as the mean fraction of the time spent in the trap. The two averages do not always coincide, as shown by the explicit computation of Section 5.2. This reveals the non-ergodic character of the cooling process, as discussed in Section 5.3.

5.1 Ensemble averages versus time averages

We define the trapped proportion $f_{\text{trap}}(\theta)$ as the probability of finding the atom in the trap at time $t = \theta$. Therefore, $f_{\text{trap}}(\theta)$ corresponds to an *ensemble average*, over many independent realizations of the stochastic process of Fig. 3.1. It is instructive to consider also a *time average*, by examining how a given atom shares its time between the 'inside' and the 'outside' of the trap. Because of the non-ergodic character of subrecoil laser cooling, ensemble averages and time averages do not in general coincide. In fact, we will see later on that the ensemble average $f_{\text{trap}}(\theta)$ and the time average only coincide when $\langle \tau \rangle$ and $\langle \hat{\tau} \rangle$ are finite, whereas they differ when either μ or $\hat{\mu}$ is smaller than one.

5.1.1 Time average: fraction of time spent in the trap

If both the average trapping time $\langle \tau \rangle$ and average return time $\langle \hat{\tau} \rangle$ are finite, the fraction of time spent by one atom in the trap is obviously given by

$$\frac{\langle \tau \rangle}{\langle \tau \rangle + \langle \hat{\tau} \rangle}. \tag{5.1}$$

If on the other hand these average times diverge, which is the case when the power-law exponents μ and $\hat{\mu}$ are both smaller than one, the previous chapter tells us that, after N visits to the trap, the total time spent inside the trap T_N grows typically as $\tau_b N^{1/\mu}$, while the total time spent outside the trap $\hat{T}_N$ grows as $\hat{\tau}_b N^{1/\hat{\mu}}$. Depending on whether μ is smaller or larger than $\hat{\mu}$, T_N will be much larger or much smaller than $\hat{T}_N$. More precisely, the fraction of time spent by the atom in the trap in this case is given by:

$$\frac{\xi \tau_b N^{1/\mu}}{\xi \tau_b N^{1/\mu} + \hat{\xi} \hat{\tau}_b N^{1/\hat{\mu}}}, \quad \text{for} \quad N \to \infty, \tag{5.2}$$

where ξ and $\hat{\xi}$ are Lévy distributed random variables of order one, with Lévy indices equal to μ and $\hat{\mu}$, respectively.

Let us suppose for definiteness that $\hat{\mu} < \mu$. In this case, the above fraction of time spent in the trap behaves, for large N, as $\tau_b N^{1/\mu}/\hat{\tau}_b N^{1/\hat{\mu}}$, implying that most of the time will be spent outside the trap. One has therefore:

$$\theta \simeq \hat{T}_N \simeq \hat{\xi} \hat{\tau}_b N^{1/\hat{\mu}}. \tag{5.3}$$

The fraction of time that a given atom typically spends within the trap thus decays with time as:

$$\frac{\xi}{\hat{\xi} \hat{\mu}/\mu} \frac{\tau_b}{\hat{\tau}_b} \left(\frac{\theta}{\hat{\tau}_b} \right)^{\hat{\mu}/\mu - 1}. \tag{5.4}$$

We can see that simply knowing the exponents μ and $\hat{\mu}$ provides information on the competition between trapping and recycling.

5.1.2 Ensemble average: trapped proportion

To compute the ensemble average $f_{\text{trap}}(\theta)$, let us come back to Fig. 3.1, which represents a random sequence of alternating trapping and escape periods. Initially, the atom is out of the trap. At time $t = \hat{\tau}_1$, it returns to the trap (point R_1). It remains in the trap during a time τ_1 and escapes from it at time $t = \hat{\tau}_1 + \tau_1$ (point E_1). Then, it diffuses out of the trap during a time $\hat{\tau}_2$, before returning to the trap at time $t = \hat{\tau}_1 + \tau_1 + \hat{\tau}_2$ (point R_2), and so on. The alternate random sequence of points $R_1, E_1, R_2, E_2, \ldots, R_i, E_i, \ldots$ gives the times at which the atom returns to the trap (points R) and the times at which it exits from the trap (points E). If we average over many independent realizations of such a random sequence, we can, as in Section 4.4, introduce a sprinkling distribution $S_R(t)$ of points R, representing the probability density of finding a point R at time t, independently of the number of previous R points. $S_R(t) \, dt$ is in fact the probability that the atom enters the trap between t and $t + dt$. Similarly, one can associate a sprinkling distribution

$S_E(t)$ of points E, $S_E(t)\,\mathrm{d}t$ being the probability that the atom escapes from the trap between t and $t + \mathrm{d}t$.

Consider now an atom which is in the trap at time $t = \theta$. This atom last entered the trap at time $t_l < \theta$, with a probability density $S_R(t_l)$, and remained there at least up to time θ, which occurs only if the trapping time τ is larger than $\theta - t_l$. One can thus write:

$$f_{\text{trap}}(\theta) = \int_0^\theta \mathrm{d}t_l\, S_R(t_l)\, \psi(\theta - t_l), \tag{5.5}$$

where

$$\psi(\tau) = \int_\tau^\infty \mathrm{d}\tau'\, P(\tau') = 1 - \int_0^\tau \mathrm{d}\tau'\, P(\tau') \tag{5.6}$$

is the probability that the atom remains in the trap for a time longer than τ. The calculation of $f_{\text{trap}}(\theta)$ thus reduces to that of $S_R(t)$ and $\psi(t)$. This is what we do in the next section, using the Laplace transforms of these functions.

5.2 Calculation of the proportion of trapped atoms

5.2.1 Laplace transforms of the sprinkling distributions associated with the return and exit times

Since the initially uncooled atomic ensemble has a relatively large momentum spread, much larger than p_{trap}, we assume that all atoms are initially untrapped[1]. We then obtain two independent relations between the sprinkling distribution $S_R(t)$ of return times and the sprinkling distribution $S_E(t)$ of exit times, $P(\tau)$ and $\hat{P}(\hat{\tau})$:

$$S_R(t) = \hat{P}(t) + \int_0^t \mathrm{d}t_l\, S_E(t_l)\hat{P}(t - t_l), \tag{5.7a}$$

$$S_E(t) = \int_0^t \mathrm{d}t_l\, S_R(t_l) P(t - t_l). \tag{5.7b}$$

Equation (5.7a) means that either the atom returns to the trap for the first time at time t, with a probability density $\hat{P}(t)$ (point R_1 of Fig. 3.1), or that it has already returned one or more times before, the last exit occurring at time t_l, with a probability density $S_E(t_l)$, and that the atom has remained out of the trap for a time $t - t_l$ before returning to the trap at time t. Equation (5.7b) means that the atom coming out of the trap has already returned to it before (one or several times), the last return occurring at time t_l, with a probability density $S_R(t_l)$, and the atom has remained in the trap for a time $(t - t_l)$ before exiting at time t. The

---

[1] The generalization of our treatment to a finite proportion of initially trapped atoms is straightforward.

formal asymmetry between eq. (5.7a) and eq. (5.7b) comes from the assumption of initially untrapped atoms.

By taking the Laplace transforms of eq. (5.7a) and eq. (5.7b), we have

$$\mathcal{L}S_R(s) = \mathcal{L}\hat{P}(s)\,[1 + \mathcal{L}S_E(s)], \tag{5.8a}$$

$$\mathcal{L}S_E(s) = \mathcal{L}P(s)\mathcal{L}S_R(s). \tag{5.8b}$$

By solving this system, we finally obtain

$$\mathcal{L}S_R(s) = \frac{\mathcal{L}\hat{P}(s)}{1 - \mathcal{L}P(s)\mathcal{L}\hat{P}(s)}, \tag{5.9a}$$

$$\mathcal{L}S_E(s) = \frac{\mathcal{L}P(s)\mathcal{L}\hat{P}(s)}{1 - \mathcal{L}P(s)\mathcal{L}\hat{P}(s)}. \tag{5.9b}$$

We can derive eq. (5.9b) in another way. Indeed, the exit points E_1, E_2, E_3, ... in Fig. 3.1 occur at times $\tilde{\tau}_1$, $\tilde{\tau}_1 + \tilde{\tau}_2$, $\tilde{\tau}_1 + \tilde{\tau}_2 + \tilde{\tau}_3$, ... where

$$\tilde{\tau}_i = \tau_i + \hat{\tau}_i. \tag{5.10}$$

The density $S_E(t)$ of exit points E_i is therefore the sprinkling distribution associated with $\widetilde{P}(\tilde{\tau})$, the probability distribution of $\tilde{\tau}$. The results established in Section 4.4 thus apply directly. Since τ and $\hat{\tau}$ are independent random variables, $\widetilde{P}(\tilde{\tau})$ is the convolution product $P(\tau) \otimes \hat{P}(\hat{\tau})$. Using eq. (4.42), one again finds eq. (5.9b).

5.2.2 Laplace transform of the proportion of trapped atoms

To calculate $\mathcal{L}f_{\text{trap}}(s)$ we take the Laplace transform of eq. (5.5) which gives:

$$\mathcal{L}f_{\text{trap}}(s) = \mathcal{L}S_R(s)\,\mathcal{L}\psi(s). \tag{5.11}$$

The Laplace transform of $\psi(\tau)$ defined in eq. (5.6) is

$$\mathcal{L}\psi(s) = \frac{1}{s} - \frac{1}{s}\mathcal{L}P(s). \tag{5.12}$$

Inserting eq. (5.12) and eq. (5.9a) into eq. (5.11) finally gives:

$$\mathcal{L}f_{\text{trap}}(s) = \frac{\mathcal{L}\hat{P}(s)}{1 - \mathcal{L}P(s)\mathcal{L}\hat{P}(s)}\,\frac{1 - \mathcal{L}P(s)}{s}. \tag{5.13}$$

Note that this expression can be rewritten as

$$\mathcal{L}f_{\text{trap}}(s) = \frac{1}{s}\frac{\mathcal{L}\hat{P}}{1 - \mathcal{L}P\mathcal{L}\hat{P}} - \frac{1}{s}\frac{\mathcal{L}P\mathcal{L}\hat{P}}{1 - \mathcal{L}P\mathcal{L}\hat{P}}, \tag{5.14}$$

where one recognizes the expressions (5.9a) and (5.9b) of $\mathcal{L}S_{\text{R}}(s)$ and $\mathcal{L}S_{\text{E}}(s)$

$$\mathcal{L}f_{\text{trap}}(s) = \frac{\mathcal{L}S_{\text{R}}(s)}{s} - \frac{\mathcal{L}S_{\text{E}}(s)}{s}, \tag{5.15}$$

from which we infer

$$f_{\text{trap}}(\theta) = \int_0^\theta [S_{\text{R}}(t) - S_{\text{E}}(t)]\,\mathrm{d}t. \tag{5.16}$$

This relation has an interesting interpretation and could have been guessed from the start. It simply states that the number of trapped atoms at time θ is equal to the number of atoms that have entered the trap between $t = 0$ and $t = \theta$ ($\int_0^\theta S_{\text{R}}(t)\,\mathrm{d}t$) minus the number of atoms that have left the trap during the same period ($\int_0^\theta S_{\text{E}}(t)\,\mathrm{d}t$).

We have thus derived an exact expression for $\mathcal{L}f_{\text{trap}}(s)$. Using the small-$s$ expansion of $\mathcal{L}P(s)$ and $\mathcal{L}\hat{P}(s)$, one can now obtain the small-s expansion of $\mathcal{L}f_{\text{trap}}(s)$, which will allow one to study the long time behaviour of $f_{\text{trap}}(\theta)$. This is what we do in the next section.

5.2.3 Results for a finite average trapping time and a finite average recycling time

Suppose first that μ and $\hat{\mu}$ are larger than one, so that $\langle\tau\rangle$ and $\langle\hat{\tau}\rangle$ are finite. In such a case, the small-s expansion of $\mathcal{L}P(s)$ (eq. (4.23)) reads:

$$\mathcal{L}P(s) = 1 - s\langle\tau\rangle - O(s^{\min(\mu,2)}) \tag{5.17}$$

and similarly for $\mathcal{L}\hat{P}(s)$. The small-s expansion of eq. (5.13) then leads to:

$$\mathcal{L}f_{\text{trap}}(s) = \frac{\langle\tau\rangle}{\langle\tau\rangle + \langle\hat{\tau}\rangle}\frac{1}{s} + O(1), \tag{5.18}$$

which shows that

$$f_{\text{trap}}(\theta) \xrightarrow[\theta\to\infty]{} \frac{\langle\tau\rangle}{\langle\tau\rangle + \langle\hat{\tau}\rangle}. \tag{5.19}$$

5.2.4 Results for an infinite average trapping time and a finite average recycling time

We consider now the case which is most relevant experimentally, i.e. $\mu < 1$ and $\hat{\mu} > 1$. We assume that the subleading term of $P(\tau)$ (i.e. the correction to

$\mu \tau_b^\mu / \tau^{1+\mu}$) decays faster than τ^{-2} when τ is large. In this case, as explained in the previous chapter, the small-s expansion of $\mathcal{L}P(s)$ reads (see eq. (4.14)):

$$\mathcal{L}P(s) = 1 - \Gamma(1 - \mu)(s\tau_b)^\mu - A_0(s\tau_b) + \cdots$$

where A_0 is a certain constant determined by the detailed shape of $P(\tau)$, in partic-ular p_{trap}. The small-s expansion of $\mathcal{L}\hat{P}(s)$ is still of the form (5.17).

It follows from eq. (5.9a) that the small-s expansion of $\mathcal{L}S_R$ (which will also be of use in Chapters 6 and 9) is given by:

$$\mathcal{L}S_R(s) = \frac{1}{\Gamma(1 - \mu)} (s\tau_b)^{-\mu} - \frac{A_0\tau_b + \langle \hat{\tau} \rangle}{[\Gamma(1 - \mu) \tau_b^\mu]^2} s^{1-2\mu} + \cdots . \tag{5.20}$$

The corresponding expression of $S_R(t)$ for large times reads, using eq. (4.24) and eq. (4.47):

$$S_R(t) = \frac{\sin(\pi\mu)}{\pi} \tau_b^{-\mu} t^{\mu-1} + \cdots . \tag{5.21}$$

For small s, using eq. (4.14) in eq. (5.12), together with eq. (5.20) and eq. (5.11), finally gives[2]:

$$\mathcal{L}f_{\text{trap}}(s) = \frac{1}{s} - \frac{\langle \hat{\tau} \rangle}{\Gamma(1 - \mu) (s\tau_b)^\mu} + \cdots \tag{5.22}$$

which shows that $f_{\text{trap}}(\theta) \to 1$ for $\theta \to \infty$. This is to be expected since the average time in the trap diverges, whereas the average time outside the trap is finite. Furthermore, eq. (5.22) allows us to determine how $f_{\text{trap}}(\theta)$ tends towards its limit when $\theta \to \infty$. Using eqs. (4.24) and (4.47) of the previous chapter, one deduces from eq. (5.22) that:

$$f_{\text{trap}}(\theta) \underset{\theta \to \infty}{=} 1 - \frac{\sin \pi\mu}{\pi} \frac{\langle \hat{\tau} \rangle}{\tau_b^\mu \theta^{1-\mu}} + \cdots . \tag{5.23}$$

We will check such a prediction in Chapter 8 by comparing it with the results of numerical simulations using the delay function.

The *time scale* θ_0 beyond which the trapped proportion begins to be significant and beyond which eq. (5.23) begins to be valid (the second term is less than one) is

$$\theta_0 \simeq \left(\frac{\langle \hat{\tau} \rangle}{\tau_b^\mu} \right)^{1/(1-\mu)} . \tag{5.24}$$

The generalized CLT provides a physical interpretation of this time scale. The derivation of eq. (5.23) required (see eq. (5.20)) the sprinkling distribution to be dominated by

---

[2] We neglect $A_0\tau_b$ in comparison to $\langle \hat{\tau} \rangle$, since τ_b is of the order of τ_{trap} which is much smaller than the average return time $\langle \hat{\tau} \rangle$ to the trap.

trapping times, i.e. it requires $T_N \gg \hat{T}_N$. Using the generalized CLT (see Section 4.2), we have $T_N \simeq \tau_b N^{1/\mu}$ and $\hat{T}_N \simeq N\langle \hat{\tau} \rangle$. Thus $T_N \gg \hat{T}_N$ is satisfied as long as $N \gg N_0$, $N_0 = (\langle \hat{\tau} \rangle / \tau_b)^{\mu/(1-\mu)}$ being defined by $T_{N_0} = \hat{T}_{N_0}$. One therefore has $T_N \gg \hat{T}_N$ as soon as $\theta \gg T_{N_0} = \hat{T}_{N_0} = (\langle \hat{\tau} \rangle / \tau_b^\mu)^{1/(1-\mu)} = \theta_0$.

5.2.5 *Results for an infinite average trapping time and an infinite average recycling time*

We finally consider the case where both μ and $\hat{\mu}$ are smaller than one, leading to infinite values for $\langle \tau \rangle$ and $\langle \hat{\tau} \rangle$. Such a situation occurs in one-dimensional VSCPT cooling in the absence of confining walls ($\hat{\mu} = 1/2$ or $\hat{\mu} = 1/4$ if the Doppler effect is taken into account; see Sections 3.4.2 and 3.4.3 and Appendix A.1).

Let us for definiteness consider the case where $\hat{\mu} < \mu$. Using eq. (4.14) and a similar equation for $\mathcal{L}\hat{P}(s)$, one can deduce from eq. (5.13) the small-s expansion of $\mathcal{L}f_{\text{trap}}(s)$ which reads:

$$\mathcal{L}f_{\text{trap}}(s) \underset{s \to 0}{\simeq} \frac{\Gamma(1-\mu)}{\Gamma(1-\hat{\mu})} \frac{\tau_b^\mu}{\hat{\tau}_b^{\hat{\mu}}} s^{\mu - \hat{\mu} - 1} + \cdots . \tag{5.25}$$

It then follows from eq. (5.25) and from eq. (4.24) that, for $\hat{\mu} < \mu$, $f_{\text{trap}}(\theta)$ decays for large θ as:

$$f_{\text{trap}}(\theta) \underset{\theta \to \infty}{\simeq} \frac{\tau_b^\mu}{\hat{\tau}_b^{\hat{\mu}}} \frac{\Gamma(1-\mu)}{\Gamma(1-\hat{\mu})\Gamma(1+\hat{\mu}-\mu)} \theta^{\hat{\mu}-\mu}, \tag{5.26}$$

with a correction term of the order of $\theta^{2(\hat{\mu}-\mu)}$ when $\mu \neq 2\hat{\mu}$ (and an extra logarithmic correction if $\mu = 2\hat{\mu}$).

The case $\mu < \hat{\mu} < 1$ is simple to understand by permuting the role of the 'trap' and the 'outside' region: in this case the fraction of trapped atoms tends to one, with corrections decaying as $\theta^{\mu-\hat{\mu}}$.

Finally, in the marginal case where $\hat{\mu} = \mu < 1$ (corresponding for instance to the one-dimensional unconfined model of Chapter 3 when $\alpha = 2$ and thus $\mu = \hat{\mu} = 1/2$), a simple calculation using eq. (5.13) and eq. (4.14) leads to:

$$f_{\text{trap}}(\theta) \underset{\theta \to \infty}{\simeq} \frac{\tau_b^\mu}{\tau_b^\mu + \hat{\tau}_b^\mu}. \tag{5.27}$$

Note that $f_{\text{trap}}(\theta)$ tends to a constant at large times (as if $\langle \tau \rangle$ and $\langle \hat{\tau} \rangle$ were finite, see Section 5.2.3) whereas the sprinkling distribution itself does *not* tend to a constant, but behaves as:

$$S_R(t) = \frac{\sin(\pi\mu)}{\pi\left(\tau_b^\mu + \hat{\tau}_b^\mu\right)} t^{\mu-1} + \cdots . \tag{5.28}$$

(We shall need this expression in Section 6.5.)

5.3 Discussion: non-ergodic behaviour of the trapped population

Let us comment on the above results in connection with the ergodicity properties of the trapping/recycling process. In the case where both $\langle \tau \rangle$ and $\langle \hat{\tau} \rangle$ are finite, the obtained result (5.19) is the ergodic result: when the average trapping and first return times are finite, the fraction of atoms at any given (large) time in the trap (ensemble average) is equal to the average fraction of the time spent in the trap by each atom individually (see eq. (5.1)). Ensemble averages and time averages then coincide.

On the other hand, when both $\langle \tau \rangle$ and $\langle \hat{\tau} \rangle$ are infinite, the fraction of atoms in the trap after a long time, given by eq. (5.26) above, does *not* coincide with the fraction of time typically spent by an atom in the trap. According to eq. (5.4), the latter decays as $\theta^{(\hat{\mu}-\mu)/\mu}$, and not as $\theta^{\hat{\mu}-\mu}$. More importantly, the prefactor of $\theta^{(\hat{\mu}-\mu)/\mu}$ depends on ξ and $\hat{\xi}$, and therefore does not cease to fluctuate from atom to atom, even in the long time limit. This clearly shows that ergodicity is broken when the average trapping time and first return time are infinite: ensemble averages no longer coincide with time averages.

Consider, for example, the case where $\mu = 1/2$ and $\hat{\mu} = 1/4$ (one-dimensional Doppler model with $\alpha = 2$, defined in Chapter 3; see also Section A.1). An individual atom typically spends in the trap a fraction of time which decays as $\theta^{-1/2}$, while the (ensemble) average fraction of atoms trapped at time $t = \theta$ only decays as $\theta^{-1/4}$. In Chapter 8 we will compare these theoretical predictions with the results of numerical simulations using the delay function and we will see that the numerical results confirm that $f_{\text{trap}}(\theta)$ decays as $\theta^{-1/4}$ and not as $\theta^{-1/2}$.

Finally, the case where $\langle \tau \rangle$ is infinite and $\langle \hat{\tau} \rangle$ finite also leads to a non-ergodic result, although the discussion is more subtle. In this case, the fraction of time spent by an atom in the trap is given by:

$$\frac{\xi \tau_b N^{1/\mu}}{\xi \tau_b N^{1/\mu} + N\langle \hat{\tau} \rangle}, \quad \text{for} \quad N \to \infty. \tag{5.29}$$

To leading order in $\theta \simeq \xi \tau_b N^{1/\mu} + N\langle \hat{\tau} \rangle \simeq \xi \tau_b N^{1/\mu}$, this leads to:

$$1 - \frac{\langle \hat{\tau} \rangle}{(\xi \tau_b)^\mu \theta^{1-\mu}}. \tag{5.30}$$

Therefore, the fraction of time spent in the trap tends to one, with a subleading correction going to zero as $\theta^{\mu-1}$, precisely as we found for the ensemble average, eq. (5.23). However, there is one crucial difference, namely that the *prefactor* of this correction term contains a random variable ξ of order one, which does not converge to a specific value, even in the long time limit. Therefore, time averages and ensemble averages do not coincide in this case either, although the difference is only detectable on the subleading corrections which vanish in the long time limit. Conceptually, however, one finds that as soon as the average trapping time diverges, the cooling process is non-ergodic.

To conclude this chapter, one can say that Lévy statistics provide quantitative predictions on the efficiency of subrecoil laser cooling. Analytical expressions have been derived for the long time limit of the proportion $f_{trap}(\theta)$ of trapped atoms. These expressions not only give the limit of $f_{trap}(\theta)$ when $\theta \rightarrow \infty$, but also the rate at which this limit is reached. Simple examples have been given of situations where ensemble averages differ from time averages, demonstrating the non-ergodic character of the cooling.

6

The momentum distribution

The fraction of *trapped* atoms $f_{\text{trap}}(\theta)$ studied in Chapter 5 gives global information on the efficiency of subrecoil laser cooling: the proportion of atoms accumulated within the sphere of radius p_{trap}. Within this sphere, one expects the momentum distribution to exhibit a narrow peak, containing the *cooled* atoms. Knowledge of $f_{\text{trap}}(\theta)$ does not provide enough information about this peak: for instance, one might trap a significant fraction of atoms within p_{trap} but, in some unfavourable cases, these atoms could be somewhat uniformly distributed over the trap, leaving only a negligible fraction in the peak itself.

In order to get a better characterization of the cooling, we calculate in this chapter the momentum distribution $\mathcal{P}(\mathbf{p})$ of the atoms contained within the sphere of radius p_{trap}. In particular, we derive analytical expressions for various important features of the narrow peak of $\mathcal{P}(\mathbf{p})$: its half-width $w(\theta)$; its height $h(\theta)$; its weight $f_{\text{peak}}(\theta)$ (which we will call the 'cooled fraction'); the shape of its tails and of its central part. We also estimate the phase space density increase associated with the cooling process. This will enable us to identify the relevant physical parameters of subrecoil cooling and to get a better understanding of the role of non-ergodicity.

6.1 Brief survey of previous heuristic arguments

Before using the statistical tools introduced in Chapter 4, which will prove to be very efficient for investigating the momentum distribution $\mathcal{P}(\mathbf{p})$, it is useful to come back to the heuristic arguments which were first used [AAK88] to estimate certain physical quantities such as the half-width $w(\theta)$ of the narrow peak. Such a discussion will show that essential features of Lévy statistics were already implicitly used in those arguments although they were not explicitly formulated.

The first estimation [AAK88] of $w(\theta)$ was done in the following way. Since only atoms of momentum p such that $R(p)\,\theta \leq 1$ can remain trapped during the whole interaction time θ, it is natural to define a time-dependent characteristic momentum

p_θ by

$$R(p_\theta)\,\theta = 1 \qquad \text{or, equivalently,} \qquad \tau(p_\theta) = \frac{1}{R(p_\theta)} = \theta \qquad (6.1)$$

which gives, using eq. (3.5):

$$p_\theta = p_0 \left(\frac{\tau_0}{\theta}\right)^{1/\alpha}. \qquad (6.2)$$

The heuristic argument consisted in conjecturing that p_θ would give the order of magnitude of the half-width $w(\theta)$ of the cooled peak:

$$w(\theta) \simeq p_\theta. \qquad (6.3)$$

Numerical solution of the Generalized Optical Bloch Equations (GOBE) for one-dimensional $\sigma^+ - \sigma^-$ laser configurations ($\alpha = 2$) indeed confirmed the $\theta^{-1/2}$ behaviour of the peak half-width predicted by eqs. (6.2) and (6.3), over the limited time range ($\simeq 1500\ \Gamma^{-1}$) reachable in a reasonable computer time [AAK89]. The order of magnitude of the prefactor in eq. (6.2) was also confirmed numerically. Two different analytical solutions of the same one-dimensional problem, based on the GOBE, agreed with eq. (6.2) and eq. (6.3) [AlK92, SSY97]. The same $\theta^{-1/2}$-dependence was also observed in a numerical solution of a specific two-dimensional laser configuration [MaA91].

Though well established in specific cases, the above heuristic argument suffers from a basic flaw which limits its generality: it makes the implicit assumption that the atoms in the cooled peak at the end of the interaction time θ did remain trapped for the whole interaction time θ. Obviously, this cannot be strictly true, since the atoms need to perform a random walk to reach the trap, which usually takes a non-negligible time. They enter the trap and they exit it several times. Those in the trap at time θ may have entered the trap the last time a short time τ before θ and their momentum p can be then much larger than p_θ because p is only restricted by the condition that such an atom must remain for a time at least equal to τ (which, in this case, is much smaller than θ). In order to be allowed to neglect the contribution of the atoms that, at time θ, have not been trapped for a very long time, *we must implicitly assume that the total time θ is actually dominated by the duration of a single event* – the trapping time for an atom having its momentum within the peak. We recognize here the unusual behaviour of Lévy statistics, where a single term can determine the behaviour of a Lévy sum. We thus expect from such a discussion that eq. (6.3) will hold approximately true only if the trapping times τ obey a broad distribution with infinite mean.

6.2 Expressions of the momentum distribution and of related quantities

6.2.1 Distribution of the momentum modulus

We first introduce the distribution $\mathcal{P}(p, \theta)$ of the momentum *modulus* p, restricted to the trapping zone $p \leq p_{\text{trap}}$. An atom trapped with momentum p at time θ might have reached the p state $(0 \leq p \leq p_{\text{trap}})$ at any time t_l $(0 \leq t_l \leq \theta)$, provided that it then remained in the trap at least until θ. The 'date' t_l is thus the *last* trapping date, i.e. it satisfies $t_l + \tau \geq \theta$ where τ is the time spent in the p state, which is distributed, conditionally to p, as $P(\tau|p)$. The probability density to reach a p state at time t_l is simply given by $\rho(p) S_R(t_l)$, where $S_R(t_l)$ is the sprinkling distribution calculated in Chapter 5 and $\rho(p) = D p^{D-1}/p_{\text{trap}}^D$ is the probability density (see eq. (3.28)) for an atom entering the trap to have a momentum modulus p (as in Chapter 3, we suppose that $p_{\text{trap}} \ll \hbar k$, so that the volume of the trap is reached uniformly). Thus the probability $\mathcal{P}(p, \theta)$ is the sum (over all possible t_l) of the probability $\rho(p) S_R(t_l)$ of reaching the trap at time t_l with momentum p, multiplied by the probability $\psi(\theta - t_l|p)$ that the trapping time τ exceeds $\theta - t_l$ for an atom with momentum p:

$$\mathcal{P}(p, \theta) = \rho(p) \int_0^\theta dt_l \, S_R(t_l) \psi(\theta - t_l|p) \tag{6.4}$$

where

$$\psi(\tau|p) = \int_\tau^\infty d\tau' P(\tau'|p). \tag{6.5}$$

Recalling that

$$P(\tau) = \int_0^{p_{\text{trap}}} dp \, \rho(p) P(\tau|p) \tag{6.6}$$

is the probability that the trapping time within p_{trap} is equal to τ, it is easy to check that, as expected:

$$\int_0^{p_{\text{trap}}} dp \, \mathcal{P}(p, \theta) = \int_0^\theta dt_l \, S_R(t_l) \int_{\theta - t_l}^\infty d\tau \, P(\tau) = f_{\text{trap}}(\theta) \tag{6.7}$$

(see eq. (5.5) and eq. (5.6)).

We shall consider in the following the two models introduced in Section 3.3.1, i.e. a deterministic model where

$$P(\tau|p) = \delta[\tau - \tau(p)] \tag{6.8}$$

and an exponential model where

$$P(\tau|p) = \frac{1}{\tau(p)} \exp\left[-\tau/\tau(p)\right]; \tag{6.9}$$

the two corresponding values of $\psi(\tau|p)$ are:

$$\psi(\tau|p) = Y[\tau(p) - \tau] \tag{6.10}$$

for the deterministic model (Y being the Heaviside function) and

$$\psi(\tau|p) = \exp[-\tau/\tau(p)] \tag{6.11}$$

for the exponential model. According to eq. (3.5), the dependence of $\tau(p)$ on the momentum p is given, in general, by:

$$\tau(p) = \tau_0 \left(\frac{p_0}{p}\right)^\alpha. \tag{6.12}$$

Finally, a general result concerning the tails of the momentum distribution can be simply derived from eq. (6.4). For $p \gg p_\theta$, we have $\tau(p) \ll \theta$ according to eq. (6.1). Equations (6.10) and (6.11) then show that $\psi(\theta - t_l|p)$ has non-zero values only if $\theta - t_l \lesssim \tau(p) \ll \theta$, so that only the region $t_l \sim \theta$ will contribute to the integral of eq. (6.4). Since $S_R(t)$ varies more slowly than $\psi(t|p)$, one can write:

$$\mathcal{P}(p, \theta) \underset{p \gg p_\theta}{\simeq} \rho(p)S_R(\theta) \int_0^\infty d\tau' \psi(\tau'|p) = \rho(p)S_R(\theta) \int_0^\infty d\tau' \tau' P(\tau'|p) \tag{6.13}$$

where the last equality follows from an integration by parts. By definition, $\tau(p)$ is the average trapping time at momentum p, and hence, we finally get the general result:

$$\mathcal{P}(p, \theta) \underset{p \gg p_\theta}{\simeq} \rho(p)S_R(\theta)\tau(p). \tag{6.14}$$

Note that the only dependence on the model is in $S_R(\theta)$ through the quantity τ_b, which is equal to τ_{trap} for the deterministic model, and to $\tau_{\text{trap}}[\mu\Gamma(\mu)]^{1/\mu}$ for the exponential model (see eq. (3.33) and eq. (3.34)). We will discuss in Chapter 7 the physical meaning of such a simple expression in terms of a 'quasi-equilibrium' regime.

6.2.2 Momentum distribution along a given axis

We suppose that the three-dimensional momentum distribution $\mathcal{P}(\mathbf{p}, \theta)$ is spherically symmetric, and we introduce the reduced momentum distribution $\pi(p, \theta)$ such that:

$$\mathcal{P}(p, \theta) = S_D p^{D-1}\pi(p, \theta), \tag{6.15}$$

where $\mathcal{P}(p, \theta)$ is the distribution of the momentum modulus introduced in Section 6.2.1 and where $S_D p^{D-1}$ is the surface of the sphere of radius p in D dimensions (see eq. (3.27)). In fact, $\pi(p, \theta)$ is the section of the three-dimensional momentum distribution $\mathcal{P}(\mathbf{p}, \theta)$ along any axis[1] passing through the origin $\mathbf{p} = \mathbf{0}$. For example,

$$\pi(p, \theta) = \mathcal{P}(p_x = p, \, p_y = 0, \, p_z = 0, \theta). \tag{6.16}$$

Equation (6.4) and eq. (3.26) then give

$$\pi(p, \theta) = \frac{1}{V_D(p_{\text{trap}})} \int_0^\theta dt_l \, S_R(t_l) \psi(\theta - t_l | p), \tag{6.17}$$

where $V_D(p_{\text{trap}}) = C_D p_{\text{trap}}^D$ is the volume of a D-dimensional sphere of radius p_{trap} (see eq. (3.24)).

In the tails ($p \gg p_\theta$), a calculation similar to the one leading to eq. (6.14) gives:

$$\pi(p, \theta) \underset{p \gg p_\theta}{\simeq} \frac{1}{V_D(p_{\text{trap}})} S_R(\theta) \, \tau(p). \tag{6.18}$$

6.2.3 Characterization of the cooled atoms' momentum distribution

From $\pi(p, \theta)$, we can define (by analogy with the rms value for a Gaussian distribution) the $e^{-1/2}$ *half-width* of the peak of cooled atoms, denoted $w(\theta)$, through:

$$\pi(p = w(\theta), \theta) = e^{-1/2} \pi(p = 0, \theta). \tag{6.19}$$

In order to characterize the momentum distribution of the trapped atoms, it is also useful to introduce the *median* momentum $p_m(\theta)$ of the trapped atoms such that:

$$\int_0^{p_m} \mathcal{P}(p, \theta) \, dp = \frac{1}{2} f_{\text{trap}}(\theta) = \frac{1}{2} \int_0^{p_{\text{trap}}} \mathcal{P}(p, \theta) \, dp. \tag{6.20}$$

The *height* $h(\theta)$ of the cooled peak is simply defined by

$$h(\theta) = \pi(p = 0, \theta) = \mathcal{P}(\mathbf{p} = \mathbf{0}, \theta). \tag{6.21}$$

From eq. (6.17), one obtains:

$$h(\theta) = \frac{1}{V_D(p_{\text{trap}})} \int_0^\theta S_R(t_l) \, dt_l \tag{6.22}$$

independently of the shape of $P(\tau|p)$, since for $p = 0$, $\tau(p) = \infty$, so that $\psi(\tau|p = 0) = 1$ (see eqs. (6.10) and (6.11)). (Note again that $S_R(t_l)$ depends on the chosen model through τ_b.)

---

[1] Note that $\pi(p, \theta)$ is *not* the probability distribution of p_x which would be obtained by integrating $\mathcal{P}(p_x, p_y, p_z, \theta)$ over p_y and p_z. The dimension of $\pi(p, \theta)$ is $1/p^D$.

Equation (6.22) can be interpreted intuitively: the height of the cooled peak is proportional to the number of atoms that have reached the state $p = 0$ between $t = 0$ and $t = \theta$. Since the probability of entering the trap between t_l and $t_l + dt$ is equal to $S_R(t_l)dt_l$, its integral indeed gives the total number of entries in the trap. The factor $1/V_D(p_{\text{trap}})$ is related to the fraction of atoms which fall in the trap at $p = 0$ (where they remain indefinitely) rather than anywhere else in the trap.

The Laplace transform of $h(\theta)$ is

$$\mathcal{L}h(s) = \frac{\mathcal{L}S_R(s)}{s\,V_D(p_{\text{trap}})},\tag{6.23}$$

an equation which will be useful later on.

One can also define the *fraction of cooled atoms* $f_{\text{peak}}(\theta)$ as the proportion of atoms of momentum less than one of the above characteristic momentum, for example p_θ:

$$f_{\text{peak}}(\theta) = \int_0^{p_\theta} \mathcal{P}(p, \theta)\,\mathrm{d}p.\tag{6.24}$$

Finally, another important physical quantity is the *phase space density* $\mathcal{D}(\theta)$ in $\mathbf{p} = 0$. In most experiments, one can neglect the increase of the spatial volume occupied by the atoms during the interaction time[2]. This is due to the fact that spatial diffusion is much slower than momentum diffusion. In such a case, the increase of the phase space density exactly reflects the increase of the momentum space density, which is described by the increase of $h(\theta)$.

The quantities $\mathcal{P}(p, \theta)$, $\pi(p, \theta)$, $h(\theta)$ and $f_{\text{peak}}(\theta)$, given by eqs. (6.4), (6.17), (6.22) and (6.24), characterize the momentum distribution. Therefore they cannot depend on the parameter p_{trap} which was introduced only for convenience in intermediate

---

[2] Note that this is not trivial since Lévy flights can also appear in real space and could lead to anomalously fast diffusion. In fact, as shown here, this is not the case because the long trapping times actually correspond to small velocities.

Let $\Pi(l)$ be the distribution of the jump lengths l in real space between two photon scatterings. If we consider only the jumps of the trapped atoms, $\Pi(l)$ is given by $\Pi(l)\mathrm{d}l = \rho(p)\mathrm{d}p$, where $\rho(p) = Dp^{D-1}/p_{\text{trap}}^D$ is the probability density for an atom entering the trap to have the momentum p. The jump length $l(p)$ for an atom with momentum p is given by the free flight relation $l(p) = p\,\tau(p)/M$ where $\tau(p) = 1/R(p) \propto p^{-\alpha}$ is the duration of the free flight for such an atom. Using such a relation between $l(p)$ and p to calculate $|\mathrm{d}p/\mathrm{d}l|$, we obtain:

$$\Pi(l) = \rho(p)\,|\mathrm{d}p/\mathrm{d}l| \propto \frac{1}{l^{1+D/(\alpha-1)}}.\tag{6.25}$$

One can recognize a broad distribution with a power-law tail described by the exponent $\mu' = D/(\alpha - 1)$. We restrict ourselves to the case $\alpha \geq 1$ where long jumps are associated with the region $p \simeq 0$. Quantitatively, for one-dimensional VSCPT with $\theta = 10^5\,\Gamma^{-1}$ and metastable helium atoms, one finds that the spatial expansion due to this Lévy flight process ($\alpha = 2$, $\mu' = 1$) is negligible compared to the expansion due to standard random walk of the untrapped atoms, which is itself negligible compared to the usual size of the cloud of atoms ($\sim$500 μm, see [Bar95]). Therefore, position diffusion can be neglected. Note finally that Lévy flights in position space can also occur in usual (not subrecoil) laser cooling [MEZ96]. Such an anomalous diffusion has been observed for a single ion trapped in an optical lattice [KSW97].

calculations. The appearance of p_{trap} in the above expressions is only formal, since all four expressions involve the product of $1/p_{\text{trap}}^{D}$ with some integral of $S_{\text{R}}(t)$, which turns out to be proportional to p_{trap}^{D}.

It is now straightforward to calculate the momentum distribution and the above related quantities. The given expressions mostly depend on the sprinkling distribution $S_{\text{R}}(t)$. The accuracy of the calculations is thus determined by the accuracy of the expression used for $S_{\text{R}}(t)$ for large t. Here, the calculations will be carried on to the leading order[3] in t. As a consequence, the results will be exact in the long time limit and approximate for intermediate times. The results depend strongly on the finiteness of $\langle \tau \rangle$ and $\langle \hat{\tau} \rangle$. We shall treat here the general case where $\mu \neq 1$. The case $\mu = 1$ – of some importance in practice – is treated in Appendix C.

6.3 Case of an infinite average trapping time and a finite average recycling time

This case is important in practice: it applies to efficient cooling schemes in which friction provides a fast recycling ($\langle \hat{\tau} \rangle$ finite) while filtering enables the accumulation of a large fraction of trapped atoms ($\langle \tau \rangle$ infinite). We have seen in Chapter 5 that the trapped fraction $f_{\text{trap}}(\theta)$ tends to one in this case.

6.3.1 Explicit form of the momentum distribution

We focus here on the momentum distribution $\pi(p, \theta)$ along a given axis and we introduce into eq. (6.17) the leading term in t of the sprinkling distribution $S_{\text{R}}(t)$ (see eq. (5.21))

$$S_{\text{R}}(t) \simeq \frac{\sin(\pi \mu)}{\pi \tau_{\text{b}}^{\mu}} t^{\mu-1}.$$

Using p_{θ} such that $\tau(p_{\theta}) = \theta$ (see eqs. (6.1) and (6.2)) and changing variables to

$$q = \frac{p}{p_{\theta}}, \tag{6.26a}$$

$$u = \frac{t_l}{\theta}, \tag{6.26b}$$

one can rewrite eq. (6.17) as:

$$\pi(p = qp_{\theta}, \theta) = \frac{\sin(\pi \mu)}{\pi C_D} \frac{\theta^{\mu}}{\tau_{\text{b}}^{\mu} p_{\text{trap}}^{D}} \int_0^1 du\, u^{\mu-1} \psi(q^{\alpha}(1 - u)) \tag{6.27}$$

---

[3] Except in Chapter 9 where the next order will be needed for optimization purposes.

where $\psi(q^\alpha(1-u))$ is equal to $Y[1-q^\alpha(1-u)]$ for the deterministic model (eq. (6.10)) and to $e^{-q^\alpha(1-u)}$ for the exponential one (eq. (6.11)). Using $\mu = D/\alpha$, $\tau_b^\mu = \mathcal{A}_\mu \tau_{\text{trap}}^\mu$ and $t_0 p_0^\alpha = \tau_{\text{trap}} p_{\text{trap}}^\alpha = \theta p_\theta^\alpha$, we can then transform eq. (6.27) into

$$\pi(p = qp_\theta, \theta) = \frac{\sin(\pi\mu)}{\pi\mu\mathcal{A}_\mu C_D} \frac{1}{p_\theta^D} \mathcal{G}(q) \tag{6.28}$$

where the function $\mathcal{G}(q)$ is equal to:

$$q \leq 1: \quad \mathcal{G}(q) = 1 \tag{6.29a}$$

$$q \geq 1: \quad \mathcal{G}(q) = 1 - \left(1 - q^{-\alpha}\right)^\mu \tag{6.29b}$$

for the deterministic model (6.10), and

$$\mathcal{G}(q) = \mu \int_0^1 du \; u^{\mu-1} e^{-(1-u)q^\alpha} \tag{6.30}$$

for the exponential model (6.11)[4]. Note that for both models, we have chosen to impose $\mathcal{G}(0) = 1$, rather than to normalize the integral of $\mathcal{G}$ to the same value. In both cases, $\mathcal{G}(q \to \infty) \simeq \mu q^{-\alpha}$ (in the latter case, only the neighbourhood of $u = 1$ contributes to the integral defining $\mathcal{G}$).

Using the fact that $\mathcal{G}(0) = 1$, expression (6.28) for $\pi(p, \theta)$ can be written in terms of the reduced momentum $q = p/p_\theta$ and the height $h(\theta)$ of the cooled peak as:

$$\pi(p, \theta) = h(\theta) \, \mathcal{G}\left(\frac{p}{p_\theta}\right) = h(\theta) \, \mathcal{G}\left[\frac{p}{p_0}\left(\frac{\theta}{\tau_0}\right)^{1/\alpha}\right] \tag{6.31}$$

with[5]

$$h(\theta) = \frac{\sin(\pi\mu)}{\pi\mu\mathcal{A}_\mu C_D} \frac{1}{p_\theta^D} \propto \theta^\mu. \tag{6.32}$$

The functions $\mathcal{G}(q)$ are drawn in Fig. 6.1 for $\alpha = 2$ and $D = 1$, corresponding to $\mu = 1/2$. The tails of $\mathcal{G}(q)$ vary as $1/(2q^2)$ in agreement with the naive 'ergodic' result $\pi(p, \theta) \propto \tau(p)$, i.e. the population of the p state is proportional to the mean residence time $\tau(p)$ in this state. However, the Lorentzian tails of $\mathcal{G}(q)$ do not imply that $\mathcal{G}(q)$ itself is a Lorentzian. In Fig. 6.2 we compare $\mathcal{G}(q)$ for the exponential model with a Lorentzian having the same normalization and the same tails as $\mathcal{G}(q)$. The important point is that $\mathcal{G}(q)$ is much 'flatter' than the Lorentzian for $q \leq 1$. This is also particularly clear for the deterministic case (see Fig. 6.1)

---

[4] The expression (6.30) for the exponential model is a confluent hypergeometric function: $\mathcal{G}(q) = M(1, 1+\mu, -q^\alpha)$ (see eq. (13.2.1) in [AbS70]).

[5] An expression of $h(\theta)$ with subleading terms is given in Section 9.4, with an interesting interpretation.

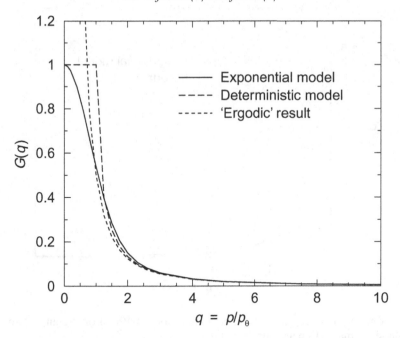

Fig. 6.1. Line shape $\mathcal{G}(q)$ with $q = p/p_\theta$ for $\alpha = 2$ and $D = 1$. The long-dashed curve corresponds to the deterministic model and the solid curve corresponds to the exponential one. The dashed curve represents the 'ergodic' result $1/(2q^2)$ having the same asymptotic behaviour for $q \to \infty$. The characteristic values q_e and q_m defined in Section 6.3.2 are, for the exponential model: $q_e = 0.890\ldots$ and $q_m = 0.798\ldots$.

since $\mathcal{G}(q)$ has a perfectly horizontal plateau for $0 \leq q \leq 1$. We will discuss in the next chapter the physical meaning of such a behaviour and relate it to the non-ergodic character of the cooling process. The difference between $\mathcal{G}$ and a Lorentzian can actually be probed experimentally – see [SLC99] and Section 8.4.3 (Fig. 8.8).

6.3.2 Important features of the momentum distribution

There are quite a number of results which can be deduced from expression (6.31).

(i) The momentum distribution $\pi(p, \theta)$ remains self-similar for any θ: $\pi(p, \theta)$ is always given by $\mathcal{G}(q)$ with a proper rescaling of the height and the width.

(ii) The auxiliary parameter p_{trap} no longer appears in expressions (6.31) or (6.32), as expected.

(iii) The momentum p_{max}, which fixes the average recycling time, is also absent from these expressions. This stems from the domination of $S_R(t)$ by trapping times τ whose distribution $P(\tau)$ is broad. Provided that $\langle \hat{\tau} \rangle$ is finite, i.e.

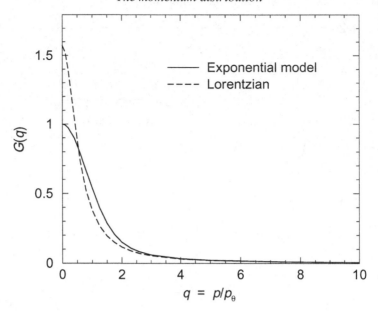

Fig. 6.2. Comparison of $\mathcal{G}(q)$ for the exponential model with a Lorentzian line shape that has the same normalization and the same tails.

provided that p_{max} is finite, the recycling times $\hat{\tau}$ play no role to leading order when θ is large. This remark also applies to all the quantities related to $\pi(p, \theta)$.

(iv) The height $h(\theta)$ increases with θ, which is *the signature of cooling*. This height increase has a power-law dependence with exponent $\mu = D/\alpha$ determined only by the long tail behaviour of $P(\tau)$.

(v) Since the only p-dependence of $\pi(p, \theta)$ is through the reduced momentum $q = p/p_\theta$, it is clear that the $1/\sqrt{e}$ half-width $w(\theta)$ is given by

$$w(\theta) = q_e p_\theta \qquad \text{with} \qquad \mathcal{G}(q_e) = \frac{1}{\sqrt{e}}. \qquad (6.33)$$

In eq. (6.33) q_e is a numerical factor which depends on μ and α. For the case $\mu = 1/2$, $\alpha = 2$, and in the exponential model, one finds $q_e = 0.890\ldots$. This width thus corresponds, up to a numerical prefactor, to the characteristic momentum p_θ (eq. (6.2)). The same is true for the median momentum $p_m(\theta)$ defined by eq. (6.20), with a different prefactor q_m, defined (for $D = 1$) as

$$\int_0^{q_m} dq \, q^{D-1} \mathcal{G}(q) = \frac{1}{2} \int_0^{q_{trap}} dq \, q^{D-1} \mathcal{G}(q) \qquad (6.34)$$

where $q_{trap} = p_{trap}/p_\theta$. For the case $\mu = 1/2$, $\alpha = 2$, and still in the exponential model, one finds $q_m = 0.79\ldots$. Thus, at any time θ, most trapped atoms are indeed characterized by a momentum of order p_θ.

(vi) The cooled fraction $f_{\text{peak}}(\theta)$ is computed from eq. (6.24) by using eqs. (6.27) and (6.15). The parameter p_θ is then eliminated thanks to eq. (6.2). One finally obtains that $f_{\text{peak}}(\theta)$ is a constant (with a value between zero and one), independent of time:

$$f_{\text{peak}}(\theta) = \frac{D \sin(\pi \mu)}{A_\mu \pi \mu} \int_0^1 dq \, q^{D-1} \mathcal{G}(q). \qquad (6.35)$$

For $D = 1$, $\mu = 1/2$, one finds $f_{\text{peak}}(\theta) = 0.59\ldots$ (for the exponential model).

(vii) It is very important to realize that, even if a finite fraction of the trapped atoms have a momentum less than the $e^{-1/2}$ half-width $w(\theta)$, the tails of the distribution are much 'fatter' than for a Maxwellian distribution with the same width. For $p_\theta \ll p \le p_{\text{trap}}$, corresponding to $\tau(p) \ll \theta$, one can use eq. (6.18) which shows that $\pi(p, \theta)$ varies as $S_R(\theta)\tau(p)$ for $p \gg p_\theta$. One can also use the asymptotic dependence $\mathcal{G}(q \to \infty) \simeq \mu q^{-\alpha}$ to obtain:

$$\pi(p, \theta) \underset{p \gg p_\theta}{\simeq} \mu h(\theta) \left(\frac{p_\theta}{p} \right)^\alpha \propto \frac{1}{\theta^{1-\mu} p^\alpha}. \qquad (6.36)$$

Thus the momentum distribution tails decay with a power-law p-dependence. In particular, for $\alpha = 2$, it decays only as p^{-2}, i.e. as a Lorentzian. The average square momentum is not of order p_θ^2 but rather of order $p_{\text{trap}}^D p_\theta^{2-D} \gg p_\theta^2$. We note, however, that in the present case $\mu < 1$, the value of $\pi(p, \theta)$ at a given momentum p decays with θ for $p \gg p_\theta$: the tails therefore shrink. The time evolution of the momentum distribution is shown in Fig. 6.3.

To sum up, the case treated in this section ($\langle \tau \rangle$ infinite and $\langle \hat{\tau} \rangle$ finite) passes all the criteria of efficient cooling: the height of the cooled peak increases with time, its weight is a significant fraction of one and the amplitude of the tails vanishes at large times. However, the shape of the peak is not Maxwell–Boltzmann, but rather has 'fat tails' and a 'flat top'!

6.4 Case of a finite average trapping time and a finite average recycling time

This case is also important in practice: it applies to schemes in which friction provides fast recycling ($\langle \hat{\tau} \rangle$ finite), while filtering is not selective enough to provide an infinite average trapping time ($\langle \tau \rangle$ finite) – which happens when the dimension of space D is larger than the filtering exponent α. In this case, as we have discussed

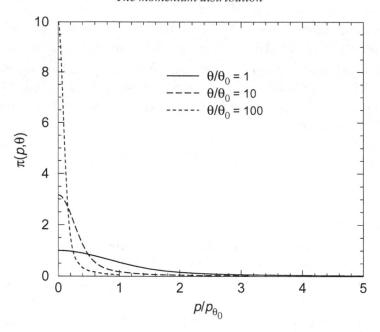

Fig. 6.3. Evolution of the momentum distribution $\pi(p, \theta)$ in the exponential model as θ increases, for $\alpha = 2$ and $D = 1$. The parameter θ_0 is an arbitrary fixed time scale, and p_{θ_0} is the corresponding characteristic momentum. Note that the distribution sharpens with time, while the amplitude of the tails decreases.

in Chapter 5, the trapped fraction $f_{\text{trap}}(\theta)$ tends at large times to a constant

$$f_{\text{trap}}(\theta) = \frac{\langle \tau \rangle}{\langle \tau \rangle + \langle \hat{\tau} \rangle} = \left(1 + (\mu - 1) \frac{p_{\text{max}}^D}{p_{\text{trap}}^{D-\alpha} p_0^{\alpha}} \right)^{-1}. \tag{6.37}$$

As this constant vanishes for $p_{\text{trap}} \to 0$, one might think that subrecoil cooling is inefficient in this case. In fact, the following calculations show unambiguously that subrecoil cooling remains efficient even in this case.

6.4.1 Explicit form of the momentum distribution

The function $S_{\text{R}}(t)$ is now given by (see eq. (3.35) and eq. (3.56))[6]:

$$S_{\text{R}}(t) \simeq \frac{1}{\langle \tau \rangle + \langle \hat{\tau} \rangle} = \left[\tau_0 \left(\frac{\mu}{\mu - 1} \left(\frac{p_0}{p_{\text{trap}}} \right)^{\alpha} + \left(\frac{p_{\text{max}}}{p_{\text{trap}}} \right)^{D} \right) \right]^{-1}. \tag{6.38}$$

---

[6] Eq. (6.38) is exact for the deterministic model. For the exponential model, the first term in the denominator involves a prefactor of order one.

In the limit where $p_{\text{max}} \gg p_{\text{trap}}, p_0$, the above expression simplifies to:

$$S_R(t) \simeq \frac{1}{\langle \hat{\tau} \rangle} = \left[\tau_0 \left(\frac{p_{\text{max}}}{p_{\text{trap}}} \right)^D \right]^{-1}. \tag{6.39}$$

Introducing this formula into eq. (6.17), one finally obtains, after simple integration:

$$\begin{aligned} \pi(p, \theta) &= \frac{1}{C_D p_{\text{max}}^D} \frac{\theta}{\tau_0} &&\text{if} &&p \le p_\theta \\ &= \frac{1}{C_D p_{\text{max}}^D} \frac{\tau(p)}{\tau_0} &&\text{if} &&p \ge p_\theta \end{aligned} \tag{6.40}$$

for the deterministic model (6.10), and

$$\pi(p, \theta) = \frac{1}{C_D p_{\text{max}}^D} \frac{\tau(p)}{\tau_0} \left[1 - \exp\left(-\frac{\theta}{\tau(p)} \right) \right] \tag{6.41}$$

for the exponential model (6.11).

These momentum distributions can also be written in a simple scaling form similar to eq. (6.31):

$$\pi(p, \theta) = h(\theta) \widetilde{\mathcal{G}} \left(\frac{p}{p_\theta} \right) = h(\theta) \widetilde{\mathcal{G}} \left[\frac{p}{p_0} \left(\frac{\theta}{\tau_0} \right)^{1/\alpha} \right] \tag{6.42}$$

with

$$h(\theta) = \frac{1}{C_D \, p_{\text{max}}^D} \frac{\theta}{\tau_0}. \tag{6.43}$$

For the deterministic model, the function $\widetilde{\mathcal{G}}(q)$ is

$$q \le 1: \quad \widetilde{\mathcal{G}}(q) = 1, \tag{6.44a}$$

$$q \ge 1: \quad \widetilde{\mathcal{G}}(q) = q^{-\alpha}. \tag{6.44b}$$

For the exponential model, it becomes

$$\widetilde{\mathcal{G}}(q) = q^{-\alpha} \left[1 - \exp(-q^\alpha) \right]. \tag{6.45}$$

Note that $\widetilde{\mathcal{G}}(0) = 1$ and that $\widetilde{\mathcal{G}}(q \to \infty) \simeq q^{-\alpha}$ for both models. The functions $\widetilde{\mathcal{G}}(q)$ are drawn in Fig. 6.4 for $\alpha = 2$ and $D = 3$, corresponding to $\mu = 3/2$. As in the previous section, the distribution $\pi(p, \theta)$ still presents a plateau-like region for $p \le p_\theta$.

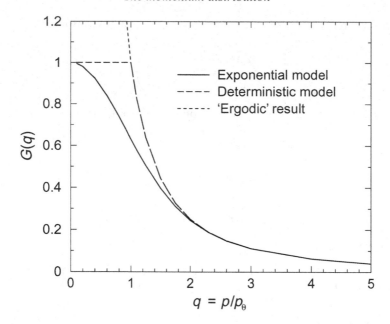

Fig. 6.4. Line shape $\widetilde{\mathcal{G}}(q)$, with $q = p/p_\theta$, for $\alpha = 2$ and $D = 3$. The long-dashed curve corresponds to the deterministic model, the solid curve to the exponential one, and the dashed curve to the 'ergodic' result $\widetilde{\mathcal{G}}(q) \propto 1/(2q^2)$ having the same behaviour as the previous curves for $q \to \infty$. The value of q_e is now 1.048, while the median momentum q_m now depends on p_{trap}.

6.4.2 Important features of the momentum distribution

We point out now a few important features of the momentum distribution.

(i) The curve $\pi(p, \theta)$ is still self-similar for any θ.

(ii) The auxiliary parameter p_{trap} is absent from the expression for $\pi(p, \theta)$.

(iii) Contrary to what happens for the case $\mu < 1$, the momentum p_{max} (which determines the average recycling time) now appears explicitly in eqs. (6.40) and (6.41). This reflects the fact that the trapping events are no longer predominant.

(iv) The height $h(\theta)$ of the peak of cooled atoms, given by eq. (6.43), increases linearly with θ. In this sense, *there is still a real cooling*. A similar increase is predicted for the phase space density.

(v) The $e^{-1/2}$ half-width q_e of $\widetilde{\mathcal{G}}(q)$ is of the order of unity, so that the $e^{-1/2}$ half-width $w(\theta) = q_e p_\theta$ of the momentum distribution $\pi(p, \theta)$ is still of the order of p_θ. However, since $\mu > 1$ is equivalent to $\alpha < D$, the integral over p of $\rho(p)\pi(p, \theta)$ is now dominated by *large* p values. The median momentum is thus given, when $p_\theta \ll p_{\text{trap}}$, by:

$$p_{\text{m}} = \frac{1}{2^{1/(D-\alpha)}} \, p_{\text{trap}}.\tag{6.46}$$

Thus, the trapped atoms are characterized in this case by a momentum of order $p_{\text{trap}} \gg p_\theta$: most trapped atoms reside on the 'border' of the trap.

(vi) The expression for the cooled fraction now reads, after changing variables to $q = p/p_\theta$:

$$f_{\text{peak}}(\theta) = Dh(\theta)p_\theta^D \int_0^1 \mathrm{d}q \, q^{D-1}\widetilde{\mathcal{G}}(q).\tag{6.47}$$

Using the fact that $h(\theta) \propto \theta$ and that $p_\theta \propto \theta^{-1/\alpha}$, one finally finds that $f_{\text{peak}}(\theta) \propto \theta^{1-\mu}$. Since now $\mu > 1$, $f_{\text{peak}}(\theta)$ decreases to 0 when $\theta \to \infty$.

(vii) It clearly appears in eq. (6.40), and also in eq. (6.41), that, for $p \gg p_\theta$, $\pi(p, \theta)$ no longer depends on θ and has a p-dependence identical to that of $\tau(p)$. Thus, as shown in Fig. 6.5, the tails of the momentum distribution reach a steady-state when θ increases. They decrease with p as a power-law $p^{-\alpha}$ (for $\alpha = 2$, the tails have a Lorentzian shape). In fact, the momentum distribution (6.40) or (6.41) remains unchanged in the tails when θ increases. Note, however, that the value p_θ of the truncation decreases when θ increases.

To sum up, the case of finite $\langle \tau \rangle$ and finite $\langle \hat{\tau} \rangle$ presents a rather subtle cooling behaviour: the cooled fraction tends to zero at large times – the trapped atoms accumulate mostly in the tails of the peak, but there is still a clear cooling effect, since the peak height and therefore the momentum and the phase space densities increase significantly.

6.5 Cases with an infinite average recycling time

The cases with $\langle \hat{\tau} \rangle$ infinite are not very favourable for cooling. Even though it seems almost always possible experimentally to make $\langle \hat{\tau} \rangle$ finite, these cases are important because several precise one-dimensional σ_+/σ_- VSCPT experiments, as well as numerical simulations, have been done with and correspond to infinite $\langle \hat{\tau} \rangle$. Moreover, one special case with infinite $\langle \hat{\tau} \rangle$ presents a significant new feature. We therefore briefly present here the various situations with infinite $\langle \hat{\tau} \rangle$ discussing in detail only the special case that brings a novel feature.

- If $\mu < \hat{\mu} \; (< 1)$, the sprinkling distribution $S_{\text{R}}(t)$ is, at first order, exactly the same as in Section 6.3, and so is the probability distribution $P(\tau)$ of the trapping times. Thus, one obtains exactly the same features for the momentum distribution as in Section 6.3, where the trapping periods also dominate.

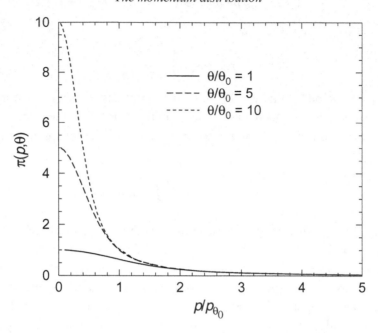

Fig. 6.5. Evolution of the momentum distribution $\pi(p, \theta)$ in the exponential model as θ increases for $\alpha = 2$, $D = 3$ ($\langle\tau\rangle$ and $\langle\hat{\tau}\rangle$ finite). The parameter θ_0 is an arbitrary fixed time scale. Note that while the height of the distribution increases with time, its tails reach a stationary (time independent) state (compare with Fig. 6.3). Note also that since $f_{\text{peak}}(\theta)$ is the integral from 0 to $p_\theta = p_0(\tau_0/\theta)^{1/\alpha}$ of the product of $\pi(p, \theta)$ and p^{D-1}, the fraction of cooled atoms goes to zero for large θ.

- If $\mu = \hat{\mu}$ (< 1), the first term of the Laplace transform of the sprinkling distribution $S_R(t)$ is the same as in the case $\mu < 1 < \hat{\mu}$ (compare eq. (5.28) and eq. (5.21)), except for the numerical prefactor. Thus, the results of Section 6.3 on the momentum distribution $\pi(p, \theta)$ still hold, except that the numerical prefactor of $\pi(p, \theta)$ will be smaller due to the finite proportion of time spent by the atoms in the recycling zone. This case applies to one-dimensional σ_+/σ_- VSCPT cooling in the regime of intermediate times for which the Doppler effect is negligible (see Sections 8.3.2, 8.4 and A.1.1.5 (p. 153)).
- If $\hat{\mu} < \mu$ (< 1), the behaviour of the cooling becomes slightly different from the previously treated cases, but the derivations are similar. This special case applies to one-dimensional σ_+/σ_- VSCPT in the long time regime for which the Doppler effect slows down the atomic diffusion at large p. The sprinkling distribution $S_R(t)$ is now dominated by recycling times and we can write, in analogy with eq. (5.21),

$$S_R(t) = \frac{\sin(\pi\hat{\mu})}{\pi} \hat{\tau}_b^{-\hat{\mu}} t^{\hat{\mu}-1} + \cdots . \qquad (6.48)$$

Proceeding as in Section 6.3.1, we obtain

$$\pi(p,\theta) = h(\theta)\,\hat{\mathcal{G}}\left(\frac{p}{p_\theta}\right) = h(\theta)\,\hat{\mathcal{G}}\left[\frac{p}{p_0}\left(\frac{\theta}{\tau_0}\right)^{1/\alpha}\right], \tag{6.49}$$

where the height $h(\theta)$ of the peak is

$$h(\theta) = \frac{\sin(\pi\,\hat{\mu})}{\pi\,\hat{\mu}C_D p_{\mathrm{trap}}^D \hat{\tau}_{\mathrm{b}}^{\hat{\mu}}}\,\theta^{\hat{\mu}} \tag{6.50}$$

and where the shape $\hat{\mathcal{G}}(q)$ is

$$q \leq 1: \qquad \hat{\mathcal{G}}(q) = 1 \tag{6.51a}$$

$$q \geq 1: \qquad \hat{\mathcal{G}}(q) = 1 - \left(1 - q^{-\alpha}\right)^{\hat{\mu}} \tag{6.51b}$$

for the deterministic model (6.10), and

$$\hat{\mathcal{G}}(q) = \hat{\mu}\int_0^1 du\, u^{\hat{\mu}-1}\mathrm{e}^{-(1-u)q^\alpha} \tag{6.52}$$

for the exponential model (6.11).

Interesting features of the momentum distribution in this case are as follows.

(i) The curve $\pi(p,\theta)$ is self-similar for any θ.
(ii) The auxiliary parameter p_{trap} is still present in $h(\theta)$ because, to maintain the generality of the treatment, we have not replaced $\hat{\tau}_{\mathrm{b}}$ by is explicit expression containing p_{trap}. If this was done, p_{trap} would disappear.
(iii) There is no momentum p_{max} in this problem.
(iv) The height $h(\theta)$ still increases in this case, in spite of the domination of recycling times over trapping times. *There is thus still real cooling.* Of course, the increase is slower than when trapping times dominate, i.e. when $\mu < \hat{\mu}$.
(v) The $\mathrm{e}^{-1/2}$ half-width q_{e} of $\hat{\mathcal{G}}(q)$ and the median q_{m} are of the order of unity. This indicates that the (few) trapped atoms are characterized by a momentum of order p_θ: most of them are in the cooled peak.
(vi) The cooled fraction $f_{\mathrm{peak}}(\theta)$ tends to zero as $\theta \to \infty$.
(vii) The tails of the momentum distribution $\pi(p,\theta)$ vary as $p^{-\alpha}$ for $p \gg p_\theta$, as in previously treated cases. On the other hand, for small momenta, we now have $\hat{\mathcal{G}}(q) \underset{q\to 0}{=} 1 - q^\alpha/(1 + \hat{\mu})$ (exponential model), while we had $\mathcal{G}(q) \underset{q\to 0}{=} 1 - q^\alpha/(1 + \mu)$ in all previously treated cases. Thus, the shape of the momentum distribution *in the vicinity of $p = 0$* now depends on $\hat{\mu}$, i.e.

it depends on the the jump rate *far away from* $p = 0$. This new feature is a remarkable sign of non-ergodicity (see Section 7.5.1).

To sum up, when $\hat{\mu} < \mu < 1$, even though the cooled fraction tends to zero at large times, there is still a clear cooling effect.

- The case $\hat{\mu} < 1$ (infinite $\langle \hat{\tau} \rangle$) and $\mu > 1$ (finite $\langle \tau \rangle$) is clearly very unfavourable: a vanishingly small fraction of atoms is in the trap, and most of them are near the border p_{trap}.

6.6 Overview of main results

In this chapter, we have derived the momentum distributions of the trapped atoms and expressed them into *scaling forms* $\mathcal{G}(q = p/p_\theta)$ that are time invariant and depend only on μ and α. All cases in which μ and $\hat{\mu}$ are different from one have been treated (the case $\mu = 1$ is treated in Appendix C). Let us concentrate here on the two most favourable cases corresponding to $\langle \hat{\tau} \rangle$ finite and $\langle \tau \rangle$ either infinite ($\mu < 1$) or finite ($\mu > 1$) (see Sections 6.3 and 6.4). The most important results are gathered in table 6.1.

Table 6.1. *Momentum distribution properties: p and θ dependence in the case where $\langle \hat{\tau} \rangle$ is finite, while $\langle \tau \rangle$ is either infinite ($\mu \leq 1$) or finite ($\mu > 1$).*

	$\mu < 1$	$\mu = 1$	$\mu > 1$
Height $h(\theta)$	θ^μ	$\theta / \log \theta$	θ
Half-width $w(\theta)$	$\theta^{-1/\alpha}$	$\theta^{-1/\alpha}$	$\theta^{-1/\alpha}$
Median p_{m}	$\simeq p_\theta \propto \theta^{-1/\alpha}$	$\simeq \sqrt{p_\theta p_{\text{trap}}} \propto \theta^{-1/2\alpha}$	$\simeq p_{\text{trap}} \propto \theta^0$
Cooled fraction $f_{\text{peak}}(\theta)$	1	$1 - \mathcal{O}((\log \theta)^{-1})$	$\theta^{1-\mu}$
Tails $\pi(p \gg p_\theta, \theta)$	$(p^\alpha \theta^{1-\mu})^{-1}$	$(p^\alpha \log \theta)^{-1}$	$(p^\alpha \theta^0)^{-1}$

In both cases ($\mu < 1$ and $\mu > 1$), the following common features were demonstrated. The height $h(\theta)$ of the cooled peak increases with θ, *which is the signature of cooling*. The half-width $w(\theta)$ of the cooled peak is proportional to p_θ. It decreases with θ as $\theta^{-1/\alpha}$ independently of the dimensionality D. The tails of the momentum distribution decay as $1/p^\alpha$.

Apart from these common features, there are important differences between the two cases $\mu < 1$ and $\mu > 1$. The parameter $p_{\max}$ does not appear in any of the characteristic momenta for $\mu < 1$, whereas it explicitly appears for $\mu > 1$ in prefactors not given in table 6.1. The cooled fraction $f_{\text{peak}}(\theta)$ tends to a constant for $\mu < 1$ whereas it decays as $1/\theta^{\mu-1}$ when θ increases for $\mu > 1$. The median momentum p_{m} is, up to a numerical prefactor, equal to p_θ when $\mu < 1$ and to p_{trap} when $\mu > 1$. Finally, the tails of the momentum distribution decrease at large times (as $1/\theta^{1-\mu}$) when $\mu < 1$, whereas they tend to a stationary value when $\mu > 1$.

We will discuss the physical content of these results in the next chapter.

7

Physical discussion

In the previous chapters, several important quantities characterizing the cooled atoms have been introduced and calculated. We now discuss the physical content of these results. We first show (Section 7.1) that the momentum distribution $\mathcal{P}(p, \theta)$ can be interpreted as the solution of a rate equation describing competition between rate of entry and rate of departure. This provides a new insight into the sprinkling distribution $S_R(t)$ which appears as a 'source term' for the trapped atoms. We then consider the tails of the momentum distribution (Section 7.2) and we show that they appear as a steady-state or 'quasi-steady'-state solution of the rate equation describing the evolution of the momentum distribution. On the contrary, in the central part of this distribution, atoms do not have the time to reach a steady-state or a quasi-steady-state because their characteristic evolution times are longer than the observation time θ. One can understand in this way the θ-dependence of the height of the peak of the cooled atoms (Section 7.3). We also investigate (Section 7.4) the important case where the jump rate $R(p)$ does not exactly vanish when $p = 0$ and we show that, when θ is increased, there is a cross-over between a regime where Lévy statistics is relevant, as in the previous case, and a regime where a true steady-state can be reached for the whole momentum distribution. We finally conclude the chapter with a few general considerations about non-stationarity, non-ergodicity and broad distributions (Section 7.5).

7.1 Equivalence with a rate equation description

7.1.1 Rate equation for the momentum distribution

In this chapter, we only consider the exponential model. Inserting eq. (6.11) into eq. (6.4) yields

$$\mathcal{P}(p, \theta) = \rho(p) \int_0^\theta dt_l \, S_R(t_l) \exp\left[-(\theta - t_l)/\tau(p)\right], \qquad (7.1)$$

where $\rho(p) = S_D p^{D-1}/V_D(p_{\text{trap}})$ is the probability density of entering the trap with momentum p (cf. eq. (3.26)). Let us now take the derivative with respect to θ of the two sides of this equation. Using the variable t instead of θ and $R(p) = 1/\tau(p)$, we get:

$$\frac{\partial}{\partial t}\mathcal{P}(p, t) = \rho(p)S_R(t) - R(p)\mathcal{P}(p, t). \tag{7.2}$$

This equation has the structure of a rate equation describing how the population of the states with momentum modulus p varies with time as a result of competition between rate of entry (the first term on the right-hand side) and rate of departure (the second term).

A similar equation can be obtained for the rate of variation of the momentum distribution $\pi(p, t)$ along a given axis. Dividing both sides of eq. (7.2) by $S_D p^{D-1}$, yields

$$\frac{\partial}{\partial t}\pi(p, t) = \frac{1}{V_D(p_{\text{trap}})}S_R(t) - R(p)\pi(p, t). \tag{7.3}$$

7.1.2 Re-interpretation of the sprinkling distribution of return times as a source term

The sprinkling distribution $S_R(t)$ plays a central role in the calculations presented in previous chapters. Up to now, we have considered it as a mean density of R points along the time axis (see Fig. 3.1), entirely determined by the distribution of the delays between two successive R points. It is in this way that we have established simple equations for the Laplace transform of $S_R(t)$ (see eq. (5.9a)).

The two equations derived above ((7.2) and (7.3)) provide a new insight into $S_R(t)$ which appears as a 'source term' for the trapped atoms, the evolution of which results from competition between rate of entry described by $S_R(t)$ and rate of departure described by the jump rate $R(p)$. We will use this point of view in the following Sections 7.2 and 7.3 to interpret the important features of the tails and the central part of the momentum distribution. Before doing such an analysis, it will be useful first to try to identify the atoms that contribute to $S_R(t)$.

7.1.3 Which atoms contribute to the sprinkling distribution of return times?

It will be simpler to identify first the atoms that contribute to the density $S_E(\theta)$ of escape times (density of E points in Fig. 3.1). As already assumed in the above analysis, we consider the case $p_{\text{trap}} \ll \hbar k$. Any trapped atom undergoing a jump therefore escapes the trap and contributes to $S_E(t)$. Since we know the jump rate $R(p)$ and the momentum distribution of the trapped atoms, for which $0 \leq p \leq$

p_{trap}, we can separately evaluate the contribution $J_{\text{peak}}(\theta)$ to $S_{\text{E}}(\theta)$ of the atoms in the peak of the momentum distribution ($0 \leq p \leq p_\theta$):

$$J_{\text{peak}}(\theta) = \int_0^{p_\theta} \mathrm{d}p\, R(p)\, \mathcal{P}(p, \theta) \tag{7.4}$$

and the contribution $J_{\text{tails}}(\theta)$ of the atoms in the tails ($p_\theta \leq p \leq p_{\text{trap}}$):

$$J_{\text{tails}}(\theta) = \int_{p_\theta}^{p_{\text{trap}}} \mathrm{d}p\, R(p)\, \mathcal{P}(p, \theta). \tag{7.5}$$

We have of course

$$S_{\text{E}}(\theta) = J_{\text{peak}}(\theta) + J_{\text{tails}}(\theta). \tag{7.6}$$

Simple calculations of the integrals of (7.4) and (7.5) then show that, in all cases, $J_{\text{peak}}(\theta)$ decreases more rapidly with θ than $J_{\text{tails}}(\theta)$ and can thus be neglected. Therefore, the atoms contributing to $S_{\text{E}}(\theta)$ are mainly those leaving the trap from the tails of the momentum distribution:

$$S_{\text{E}}(\theta) \simeq J_{\text{tails}}(\theta). \tag{7.7}$$

Let us now come back to $S_{\text{R}}(\theta)$. In the limit of long interaction times, $S_{\text{R}}(\theta)$ and $S_{\text{E}}(\theta)$ are nearly equal, since we know (see eq. (5.16)) that the integral of $S_{\text{R}} - S_{\text{E}}$ from 0 to θ is equal to $f_{\text{trap}}(\theta)$, which is at most equal to one. The flux of atoms leaving the trap is then equal to the flux of atoms coming back. We thus conclude from the above analysis that the source term $S_{\text{R}}(\theta)$ comes essentially from the atoms which leave the trap from the tails of the momentum distribution and then return afterwards to the trap.

7.1.4 *Interpretation of the time dependence of the sprinkling distribution of return times*

The conclusions of the previous subsection allow us to establish a connection between the asymptotic behaviour of the tails of the momentum distribution and the time dependence of $S_{\text{R}}(\theta)$.

As in all of this chapter, we assume that $\hat{\mu} > 1$, so that the average return time $\langle \hat{\tau} \rangle$ is finite. Suppose first that $\mu < 1$, so that the average trapping time diverges. We know from the results of Chapter 6 that the cooling is very efficient, resulting in an accumulation of the atoms in quasi-dark states with a very low jump rate. The jump rate is only due to the atoms in the tails, and since the tails shrink when θ increases, the total number of jumps per unit time decreases with θ, as well as the corresponding rate of entry into the trap. In other words, *the very fact of cooling most atoms slows down the diffusion process*. It thus appears that the decrease of

$S_R(\theta)$ with θ when $\mu < 1$ is related to the efficiency of cooling and to the shrinking of the tails of the momentum distribution.

When $\mu > 1$, the results of Chapter 6 show that the atoms in the tails of the momentum distribution reach a *steady-state* in the trap ($p_\theta < p < p_{trap}$) as well as outside the trap ($p > p_{trap}$). These atoms then form an approximately *constant reservoir* which provides a constant jump rate, and consequently a constant rate of entry into the trap. This explains why $S_R(\theta)$ becomes asymptotically independent of θ when $\mu > 1$.

7.2 Tails of the momentum distribution

7.2.1 Steady-state versus quasi-steady-state

We established in Chapter 6 a very simple expression for the tails of the momentum distribution $\mathcal{P}(p, \theta)$, corresponding to $p \gg p_\theta$ (see eq. (6.14)):

$$\mathcal{P}(p, \theta) \simeq \frac{\rho(p) S_R(\theta)}{R(p)} \quad \text{for} \quad p \gg p_\theta. \tag{7.8}$$

Such a result can be simply derived from the rate equation (7.2).

Suppose first that $\mu > 1$, so that $S_R(t)$ does not depend on t: $S_R(t) = S_R$. Equation (7.2) then admits a steady-state solution for $t \gg \tau(p) = R(p)^{-1}$. At time θ, such a steady-state regime will be reached if $\theta \gg \tau(p)$, i.e. if $p \gg p_\theta$. This condition defines the tails of the momentum distribution. For $\mu > 1$, the tails of the momentum distribution thus reach a *true* steady-state regime given by:

$$\mathcal{P}_{st}(p) = \frac{\rho(p) S_R}{R(p)} \quad \text{for} \quad p \gg p_\theta \quad \text{and} \quad \mu > 1. \tag{7.9}$$

If $\mu < 1$, the source term $\rho(p) S_R(t)$ is no longer time-independent, but varies with t as a power law: $S_R(t) \propto t^{\mu-1}$. The characteristic rate of variation of the source term is of the order of $|\dot{S}_R(t)/S_R(t)| = (1 - \mu)/t$. This means that, at time θ, the source term varies with time scales of the order of $\theta/(1 - \mu)$. If $\theta \gg \tau(p)$, i.e. in the tails of the momentum distribution, eq. (7.2) admits a 'quasi-steady-state' solution given by eq. (7.8). In other words, the power-law variations of $\rho(p) S_R(\theta)$ are slow enough, in the tails of the momentum distribution, to allow $\mathcal{P}(p, \theta)$ to *follow adiabatically* these slow variations of the source term

$$\mathcal{P}_{qst}(p, \theta) \simeq \frac{\rho(p) S_R(\theta)}{R(p)} \quad \text{for} \quad p \gg p_\theta \quad \text{and} \quad \mu < 1. \tag{7.10}$$

Thus, expression (7.8) can be interpreted as the steady-state ($\mu > 1$) or quasi-steady-state ($\mu < 1$) solution of the rate equation (7.2).

7.2.2 Dependence on the various parameters

(i) *p-dependence.* Along a given axis, the momentum distribution $\pi(p, \theta)$ is $\mathcal{P}(p, \theta)/S_D p^{D-1} \propto \mathcal{P}(p, \theta)/\rho(p)$ (see eq. (6.15)). The *p*-dependence of the tails of $\pi(p, \theta)$ is thus, according to eqs. (7.9) and (7.10), entirely determined by $1/R(p) \propto 1/p^\alpha$. It only depends on the exponent α describing the increase of $R(p)$ with p near $p = 0$. The interpretation of eq. (7.9) and eq. (7.10) as a steady-state or quasi-steady-state solution of eq. (7.2) makes such a result very clear: the shorter the departure time from a state p, the smaller the population of this state.

(ii) *θ-dependence.* The θ-dependence is entirely contained in $S_R(\theta)$, so that the tails of the momentum distribution vary as $\theta^{\mu-1}$ if $\mu < 1$, and are θ-independent if $\mu > 1$. Such a proportionality between the tails of the momentum distribution and $S_R(\theta)$ is another way of expressing the result derived in Section 7.1.3, according to which the mean number of jumps per unit time comes essentially from the atoms in the tails of the momentum distribution. This θ-dependence comes from the steady-state or quasi-steady-state character of $\mathcal{P}(p, \theta)$ for $p \gg p_\theta$.

7.3 Height of the peak of the momentum distribution

The central part of the momentum distribution corresponds to values of p such that $p < p_\theta$. The characteristic departure time, $R(p)^{-1}$, is then longer than the observation time θ, and the atoms do not have the time to reach a steady-state or quasi-steady state regime, contrary to what happens in the tails. We will discuss in Section 7.5 the consequences of such a situation in terms of non-stationarity and non-ergodicity. Here, we focus on the peak of the momentum distribution corresponding to $p = 0$.

For $p = 0$, one has $R(p) = 0$ and thus the last term of eq. (7.3) vanishes. A straightforward integration of this equation is then possible, giving:

$$h(\theta) = \frac{1}{V_D(p_{\text{trap}})} \int_0^\theta S_R(t)\, dt, \qquad (7.11)$$

which is the same as expression (6.22) obtained in Chapter 6. This equation describes how the state $p = 0$ fills up without any leak under the only effect of the source term $S_R(t)$. Equation (7.11) then simply counts the average number of entries in the trap at $p = 0$ (see Section 6.2.3).

When $\mu > 1$, $S_R(t)$ tends rapidly to a constant value S_R and the integral of eq. (7.11) grows linearly with θ, so that:

$$h(\theta) \simeq \frac{1}{V_D(p_{\text{trap}})} \int_0^{\theta} S_R \, dt = \frac{1}{V_D(p_{\text{trap}})} S_R \, \theta. \tag{7.12}$$

Since only the atoms in the tails contribute to $S_R(t)$ (see Section 7.1.3), and since these tails have reached a steady-state, one gets a simple understanding of the linear increase of $h(\theta)$ with θ when $\mu > 1$. It stems from a *constant feeding* of the state $p = 0$ by jumps coming from the almost *constant reservoir* of uncooled atoms.

When $\mu < 1$, we must use the power-law dependence $S_R(t) \propto t^{\mu-1}$, which gives

$$h(\theta) \propto \int_0^{\theta} S_R(t) \, dt \propto \frac{\theta^{\mu}}{\mu} \propto \theta \, S_R(\theta). \tag{7.13}$$

The θ-dependence of $h(\theta)$ thus appears as a product of a linear term, θ, by the slowly decreasing source term, $S_R(\theta) \propto \theta^{\mu-1}$. It is the decrease of $S_R(\theta)$ with θ which explains why $h(\theta)$ increases more slowly at large θ when $\mu < 1$ than when $\mu > 1$. At first sight, such a result seems paradoxical since all the previous discussions have shown that the situation $\mu < 1$ leads to a much more efficient cooling than when $\mu > 1$. The solution of this paradox is that the cooling is so efficient for $\mu < 1$ that it empties the reservoir of atoms which could, by jumping, feed the $p = 0$ state. Nearly all atoms are in dark states which do not fluoresce and the number of atoms which can still jump and contribute to $S_R(\theta)$ decreases.

7.4 Effect of a non-vanishing jump rate at zero momentum

We suppose now that, due to dissipative processes, the jump rate $R(p)$ no longer vanishes at $p = 0$:

$$R(p = 0) = R_0 \neq 0. \tag{7.14}$$

Such a situation is in general the result of experimental defects such as stray magnetic fields which contaminate the dark state with other states that can absorb the laser light, off-resonant excitation by the laser pulses used in Raman cooling, multiple scattering, etc. Such a residual jump rate R_0 is also intrinsically present in some cooling configurations [HLO00]. We investigate in this section to what extent the results derived in this book with $R_0 = 0$ remain valid when $R_0 \neq 0$. We will restrict ourselves to the case where the jump rate can be written[1]:

$$R(p) = R_0 + \frac{1}{\tau_0} \left(\frac{p}{p_0} \right)^{\alpha}. \tag{7.15}$$

---

[1] We assume here that R_0 does not correspond to a loss of atoms. The corresponding atoms continue to interact with the laser light and can be recycled. Other dissipative processes could also be considered, leading to a definitive loss of atoms. One can show that the results derived here when $R_0 \ll 1/\theta$ remain valid in this case too.

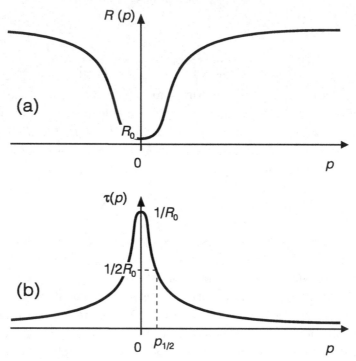

Fig. 7.1. (a) Jump rate $R(p)$ of Section 7.4. It does not vanish at $p = 0$ ($R(0) = R_0$). The increase of $R(p) - R_0$ with p is described by a power law of exponent α. (b) Trapping time $\tau(p) = 1/R(p)$. It exhibits a resonance around $p = 0$ (HWHM: $p_{1/2}$). The steady-state momentum distribution $\pi_{\text{st}}(p)$ is proportional to $\tau(p)$ and exhibits the same resonant behaviour. If R_0 is sufficiently small, $p_{1/2}$ can be much smaller than the single photon momentum $\hbar k$.

The variations with p of $R(p)$ are shown in Fig. 7.1a.

7.4.1 Existence of a steady-state for long times

All atomic evolution times are now shorter than R_0^{-1}. It follows that, after a long enough interaction time, $\theta \gg R_0^{-1}$, the system reaches a steady-state in which the sprinkling distribution $S_R(\theta)$ equals a constant S_R.

We still define the trapping zone by $p < p_{\text{trap}} < \hbar k$ so that the probability density that an atom entering the trap reaches the momentum $\mathbf{p}$ is uniform. The momentum distribution $\mathcal{P}(p, \theta)$ can be derived from the general equation (6.4) which is of course still valid here. It will be simpler to use eq. (7.2) where $S_R(\theta)$ is replaced by the constant S_R. The steady-state solution of this equation is:

$$\mathcal{P}(p, \theta) = \rho(p) S_R \, \tau(p). \tag{7.16}$$

This expression is identical to eq. (7.9) that was valid for the tails only and for $\mu > 1$. Now, with $R_0 \neq 0$, it is valid for *all* trapped atoms, including the peak.

From eq. (6.15) and eq. (3.26), one easily finds the momentum distribution along a given axis:

$$\pi(p, \theta) = \frac{1}{V_D(p_{\text{trap}})} S_R \tau(p). \tag{7.17}$$

Figure 7.1b represents the variations of $\tau(p)$ with p. The trapping time $\tau(p)$ decreases from a maximum value $\tau(p = 0) = 1/R_0$ and is divided by two when $p = p_{1/2}$, $p_{1/2}$ being such that[2]:

$$R_0 = \frac{1}{\tau_0} \left(\frac{p_{1/2}}{p_0} \right)^\alpha. \tag{7.18}$$

When $R_0 \to 0$, $p_{1/2} \to 0$. If R_0 is sufficiently small, the half-width at half-maximum $p_{1/2}$ of the momentum distribution can be much smaller than $\hbar k$. This shows that $R_0 = 0$ is not necessary for getting *subrecoil* cooling. To simplify the following discussion, we will assume here that:

$$p_{1/2} \ll p_{\text{trap}} < \hbar k. \tag{7.19}$$

7.4.2 Intermediate times

We suppose now that

$$\tau_{\text{trap}} \ll \theta \ll R_0^{-1}. \tag{7.20}$$

We still define, as in Chapter 6, a characteristic width p_θ by

$$\frac{1}{\theta} = \frac{1}{\tau_0} \left(\frac{p_\theta}{p_0} \right)^\alpha. \tag{7.21}$$

Comparing eq. (7.18) and eq. (7.21), we get, using eq. (7.20):

$$p_{1/2} \ll p_\theta \ll p_{\text{trap}}. \tag{7.22}$$

The momentum distribution is still given by eq. (7.1), where $R(p) = \tau(p)^{-1}$ is now given by eq. (7.15) and where the sprinkling distribution $S_R(t)$ must be recalculated with $R_0 \neq 0$. We want to show here with simple arguments that, when condition (7.22) is fulfilled, the results of the calculations of previous sections are not modified if one neglects R_0.

---

[2] It is the half-width $p_{1/2}$ of $\tau(p)$ that gives the half-width of the momentum peak. Note that this half-width $p_{1/2}$ can be much smaller than the half-width ($\simeq p_0$) of the dark resonance of $R(p) = 1/\tau(p)$. In early studies of laser cooling [MiR85], the intuitive belief that the momentum width $p_{1/2}$ or p_θ was given by the width p_0 of the dark resonance prevented the discovery of the subrecoil properties of Velocity Selective Coherent Population Trapping.

Consider first the exponential $\exp[-R(p)(\theta - t_l)]$ of eq. (7.1). Since t_l varies from 0 to θ, the contribution of R_0 to the argument of the exponential is at most equal to $R_0\,\theta$, which is negligible compared to one according to eq. (7.20). This is related to the fact that atomic evolution times much longer than the observation time θ cannot be distinguished. The dissipative processes introduce an atomic evolution time, R_0^{-1}, which is too long to produce an observable effect during the much shorter time θ.

To evaluate the corrections of $S_R(\theta)$ due to R_0, one can compare the number of jumps per unit time $J_0 = R_0$ due to the dissipative processes to those calculated in Section 7.1.3 with $R_0 = 0$, J_{peak} and J_{tails}. One easily finds that J_0 is negligible compared to J_{peak} and J_{tails}.

In conclusion, for intermediate interaction times θ satisfying eq. (7.20), the dissipative term $R_0 \neq 0$ introduces negligible corrections in the expression of $S_R(\theta)$ obtained with $R_0 = 0$. Therefore, the calculations of the previous sections done with $R_0 = 0$ remain valid even if $R_0 \neq 0$. All the non-ergodic features which we have previously described remain unchanged. When θ increases from 0 to ∞, we thus have a cross-over between power-law behaviours described by Lévy statistics and a steady-state regime which is reached exponentially. Of course, the difference with more trivial processes which reach a steady-state after a transient regime only makes sense if there is a large separation of time scales between the microscopic time τ_0 and the ergodic time scale (here R_0^{-1}). To observe the initial Lévy statistics regime clearly, it is therefore necessary to have a sufficiently small residual jump rate R_0.

7.5 Non-stationarity and non-ergodicity

7.5.1 *Flatness of the momentum distribution around zero momentum*

We come back to the case where $R(p) = 0$ and to the central part of the momentum distribution $\pi(p, \theta)$. Figures 6.1 and 6.4 of Chapter 6 give the shape of $\pi(p, \theta)$ around $p = 0$. In the deterministic model where the trapping time has a well defined value depending only on p, $\pi(p, \theta)$ exhibits a perfectly horizontal plateau for $0 < p < p_\theta$. If the trapping time is exponentially distributed, the plateau is no longer perfect, but the shape of $\pi(p, \theta)$ still exhibits a clear flatness around $p = 0$. Figure 6.2 shows, for example, that $\pi(p, \theta)$ is flatter than the normalized Lorentzian having the same tails for $p \gg p_\theta$.

The flatness of the momentum distribution around $p = 0$ is actually related to the fact that all momentum states $p \leq p_\theta$ have characteristic times longer than the observation time θ. In a sense, they cannot be discriminated by the cooling process, which does not last long enough. For two different values of p smaller than p_θ, the last term of eq. (7.3) cannot give rise to very different values of $\pi(p, \theta)$. Such a flatness is thus a direct manifestation of the *non-stationarity* associated with the existence of atomic evolution times longer than the observation time.

The fact that $\pi(p, \theta)$ deviates from the $1/R(p)$ law (see Fig. 6.1), which could be obtained by an extrapolation of the p-dependence of the tails described by eqs. (7.9) and (7.10) shows also that the population of a p state near $p = 0$ is not proportional to the mean time $1/R(p)$ that an atom can spend in this state. This population cannot be derived as a time average. This is a clear signature of the *non-ergodicity* of the cooling process. An experimental observation of the exact shape of $\pi(p, \theta)$ around $p = 0$ can thus be considered as a good test of the non-ergodic character of subrecoil cooling [SLC99] (see Section 8.4.3 and in particular Fig. 8.8).

7.5.2 Various degrees of non-ergodicity

The results of the calculations presented in the previous chapters clearly show that there are various degrees of non-ergodicity. Consider first the case where the mean trapping time $\langle \tau \rangle$ is infinite ($\mu < 1$). For sufficiently long interaction times, the proportion of cooled atoms is very large, and most atoms have their momentum falling in the central part of the momentum distribution where the population of a given p state cannot be derived as a time average. Most atoms exhibit then a non-ergodic behaviour and such a situation can be considered as reflecting *global non-ergodicity*.

By contrast, when $\langle \tau \rangle$ is finite ($\mu > 1$), the central part of the momentum distribution involves a vanishingly small fraction of atoms, which tends to zero when the interaction time tends to infinity. The non-ergodic behaviour of the cooling process is then exhibited by a very small fraction of atoms and such a situation can be called *fraction-limited non-ergodicity*.

7.5.3 Connection with broad distributions

It clearly appears from the previous discussion that *the broader the distribution of trapping times, the stronger the non-ergodicity*. We now try to analyse in more detail the role played by broad distributions in non-ergodic cooling. In order to understand the behaviour of the mean number of atoms trapped with a momentum p at time θ, we consider an ensemble of stochastic realizations of the random walk performed by an atom and we try to understand what kind of results one expects when averaging over an ensemble of such 'histories'. It will then appear that the peculiar properties of Lévy statistics described in Chapter 4 are very useful for obtaining new insights into the connections that exist between non-ergodic cooling and broad distributions.

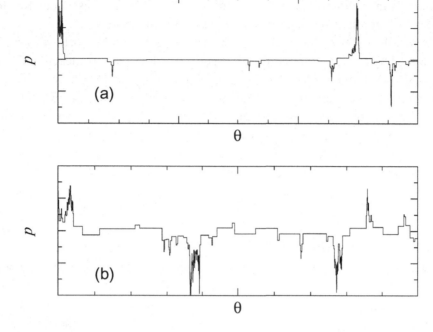

Fig. 7.2. Stochastic realizations of the random walk of the atom for $\mu = 0.5 < 1$. (a) A typical non-ergodic history, dominated by a few events. (b) An exceptional history where the longest trapping time does not exceed a tenth of the observation time θ.

The case $\mu < 1$, leading to a global non-ergodicity, corresponds to very broad distributions, so broad that $\langle \tau \rangle$ is infinite. We then know from the results of Section 4.3 that the random sequence of trapping times is dominated by a very small number of terms which are of the order of the total trapping time, which is itself of the order of the total time θ, since we suppose here that $\hat{\mu} > 1$, so that the total recycling time is negligible compared to the total trapping time. When we pick at random one stochastic realization of the random walk, the atom has a very large probability of being, at time θ, in one of these long trapping time events (see Fig. 7.2a). For most atoms, the cooling process creates time scales of the order of the total time θ. Averaging over time this ever-evolving process makes little sense and it is not surprising that time averages and ensemble averages can then differ appreciably.

The case $\mu > 1$, leading to a fraction-limited non-ergodicity, corresponds to moderately broad distributions, for which $\langle \tau \rangle$ is finite. In this case, there are still very long trapping times in the problem, those corresponding to $p < p_\theta$, but the fraction $f_{\text{peak}}(\theta)$ of cooled atoms is limited and becomes vanishingly small when θ

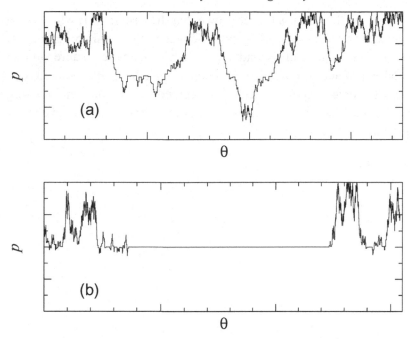

Fig. 7.3. Stochastic realizations of the random walk of the atom for $\mu = 1.5 > 1$. (a) A typical ergodic history, in which no single event dominates the history. (b) An exceptional history where the longest trapping time happens to exceed a third of the observation time θ.

increases. Such a cooling regime has now both ergodic and non-ergodic features. Again, let us pick at random one stochastic realization of the random walk. In most cases, since $f_{\text{peak}}(\theta)$ is vanishingly small, the atom has never been trapped in states of lifetimes close to θ (see Fig. 7.3a). Its history looks like an ergodic history in which no single event dominates. Most atoms are thus found at time θ in momentum distribution zones (tails of the peak and untrapped region) that tend to a stationary distribution. For these atoms, time averaging and ensemble averaging are equivalent. As they increasingly dominate the momentum distribution at long times, this cooling regime appears ergodic. However, the behaviour regarding ergodicity is more subtle here than in the case $\mu < 1$. Indeed, there are still atoms whose histories (see Fig. 7.3b) are dominated by a single event as in Fig. 7.2a. These cooled atoms represent a vanishingly small fraction but this fraction *vanishes only as a power law at large times*: $f_{\text{peak}}(\theta) \propto \theta^{-(\mu-1)}$. Thus, the cooled atoms constitute a vanishingly small fraction which plays a crucial role by ensuring that the cooling process is still efficient at long times, in the sense that the peak height $h(\theta)$ still increases.

In conclusion, this discussion has shown that the case $\mu < 1$, which corresponds to a very efficient subrecoil cooling, is associated with global non-ergodicity, while the case $\mu > 1$, which corresponds to a much less efficient subrecoil cooling, is associated with fraction-limited non-ergodicity. We have thus pointed out the link between the subrecoil cooling methods presented in this book, and non-ergodicity (as well as broad distributions). This justifies the name non-ergodic cooling sometimes given to subrecoil cooling.

8

Tests of the statistical approach

8.1 Motivation

As explained in Chapter 2, the Lévy statistics treatment of subrecoil laser cooling has been introduced after a series of simplifications, where we have dropped details of the quantum microscopic description to only keep the main features of the physical process. Such a way of reasoning is standard in statistical physics. It is difficult, however, to be sure *a priori* that one has not dropped important features, and the validity of the statistical approach needs to be checked. An important step of this verification, although not a rigorous proof, is to compare *a posteriori* the predictions of the statistical approach with experimental results as well as with the predictions of microscopic theoretical approaches. This chapter presents such comparisons.

We present in Section 8.2 the approaches (theoretical and experimental) to which our statistical approach can be compared. We then proceed to compare the results obtained by the different approaches. First, in Section 8.3, we treat in detail the predictions for a global quantity, the proportion of trapped atoms. This is done for the three recycling models introduced in this work, in the one-dimensional case. Then, in Section 8.4, we study another physical quantity with a richer content, the momentum distribution of cooled atoms. In Section 8.5, we investigate the influence of the dimensionality of the problem, and the role of friction during the recycling periods – which are crucial predictions of the Lévy statistics analysis. Finally, Section 8.6 summarizes the conclusions of the tests of the statistical approach.

To perform the comparisons between our approach and other treatments, it has been necessary to establish the correspondence between the parameters appearing in the statistical results (τ_b and $\hat{\tau}_b$), and the quantum optics parameters (atomic constants, laser detuning and intensity) characterizing the physical situation. This correspondence is given in Appendix A.

Note that this chapter does not claim to be exhaustive or to quote all the results existing on subrecoil cooling. It rather presents a few results which we think are particularly relevant for the validity of our statistical approach to subrecoil cooling.

8.2 Overview of other approaches

The most satisfactory way to test the correctness of our statistical predictions is to compare them with *experimental results* (see Section 8.2.1). However, since up to now only a limited amount of experimental data on subrecoil cooling are available, there are several predictions that have not yet been quantitatively tested by experiments.

Therefore, we also confront our statistical results with theoretical methods based on *microscopic quantum optics* calculations, resulting from the detailed description of the atom–laser interaction. The two available types of quantum optics calculations which apply to one-dimensional VSCPT are described in Section 8.2.2. Simple Monte Carlo simulations can also be performed for Raman subrecoil cooling. They are presented in Section 8.2.3. As explained below, each of these approaches has its specific advantages and scope.

8.2.1 Experiments

Since its discovery, subrecoil cooling has been a domain where experiment and theory have been strongly linked together. In the case of VSCPT, most experiments have been done with metastable helium interacting with lasers at 1.08 μm, resonant with the $2\,^3S_1 \rightarrow 2\,^3P_1$ transition [AAK88, Bar95, BSL94, LBS94, LKS95, KSP97, SHK97, Sau98, SLC99]. Some results have also been obtained with rubidium [ESW96]. It is worth noting that data on subrecoil cooling exist not only in one dimension, but also in two and three dimensions. The width of the cooled peak has been measured as a function of the cooling time θ, and the efficiency of the cooling process (in terms of velocity space density) has been investigated as a function of the dimensionality and of the presence or absence of friction. The shape of the velocity distribution of the cooled atoms has also been obtained.

In the case of Raman subrecoil cooling, experiments have been performed on alkali atoms [KaC92, DLK94, RBB95, Rei96, LAK96], and cooling has been investigated in one, two and three dimensions. A remarkable feature of these experiments for testing our statistical approach is the possibility of varying the exponent α of the power law determining the shape of the jump rate around the origin (eq. (3.5)) and therefore the tail of the trapping times distribution. This has allowed us to make a very direct test of some of the most important predictions of the Lévy statistics analysis.

Although the experimental investigations are far from being exhaustive, and although some of the data give qualitative rather than quantitative results, the importance of the available experimental results should not be underestimated. In particular, they allow one to explore the field of three-dimensional cooling, which is still out of reach of quantum optics theoretical methods and where experiment is therefore the only way to test Lévy statistics predictions.

8.2.2 *Quantum optics calculations for VSCPT*

There are two main quantum optics approaches for rigorous study of laser cooling problems, the Generalized Optical Bloch Equations (GOBE) and quantum jump simulations.

The GOBE have been used in the first study of subrecoil cooling, with one-dimensional σ_+/σ_- VSCPT [AAK89]. Starting from the effective Hamiltonian for a family of states with a given momentum p (see Appendix A), one has to solve the time-dependent master equation (GOBE) coupling the different p families.

This can first be done *numerically* and has been extremely useful in the first investigations of one-dimensional σ_+/σ_- VSCPT [AAK89] as well as for two-dimensional VSCPT schemes [MaA91, MaA92]. Unfortunately, this numerical method is not powerful enough (see Section 2.3.2) to reach the long time domain required to test our statistical analysis. In practice, we have not used it to explore interaction times longer than $10^3\Gamma^{-1}$. However, for our purposes, the numerical solution of the GOBE is valuable not only for comparison with experiments, but also as a test at short time scales of the quantum jump simulations that are then used to reach long time scales (see below). Excellent agreement is found between the two methods [Bar95, FZA95], which is to be expected since they are in principle equivalent.

The GOBE can also be solved *analytically* in the specific case of one-dimensional σ_+/σ_- VSCPT for asymptotically large times. This was first achieved in the pioneering work of Alekseev and Krylova [AlK92, AlK93]. Note that this first work omitted the Doppler effect (see note [5] in [BBE94]) and is therefore relevant to test the unconfined recycling model. The Doppler effect was included in a further work [AlK96] which enables the Doppler recycling model to be tested. Similar analytic methods have been used to study the case when one-dimensional σ_+/σ_- VSCPT is not perfect because of a relaxation term preventing perfect dark states to exist [MaM93, MKG94, Ari96]. (Note that several numerical studies [FZA95, WHF95, GPS95, MZL96] also address the question of leaks in the trapping state.) More recently, Schaufler *et al.* [SSY97, Sch98] have been able, using kinetic equations derived from GOBE, to obtain scaling laws that determine the asymptotic behaviour for one-dimensional σ_+/σ_- VSCPT. These authors make clear the connection between their results and the Lévy statistics approach.

The second main rigorous approach to laser cooling problems uses quantum jump simulations. We have used this approach extensively to study in great detail one-dimensional σ_+/σ_- VSCPT [Coh90, CBA91, Bar95, BBE94] at long times and other authors have used it for the same cooling problem [FZA95, WHF95, MZL96], for one-dimensional VSCPT schemes with friction [SHP93, WEO94, MDT94], for one-dimensional transient VSCPT [PMA92] and for a two-dimensional VSCPT scheme [WHF95]. Within this method (Section 2.3.3), the atomic evolution is described as a succession of coherent evolution periods, characterized by a constant generalized momentum p along the axis of cooling, separated by quantum jumps towards another momentum p', occurring at random times. The sequence of p values, and the dates at which p changes, can be obtained by a quantum jump procedure based on the delay function $W(\tau)$ [Coh90, CBA91].

> Thanks to this method, the computing time for one atomic history is quite modest: to reach an interaction time θ of $10^6\Gamma^{-1}$, for instance, at a Rabi frequency[1] $\Omega_1 = 0.3\Gamma$ and a null detuning, it takes less than 10 s on a personal computer [Bar95] ($\simeq 4 \times 10^7$ multiplications).

More fundamentally, this quantum jump method based on the delay function is remarkably well suited to situations in which the delays present a broad distribution. Indeed, it takes a single algorithmic step to propagate an atom over a duration of one delay, *regardless of its length* [Bar95, WHF95]. This is a crucial asset for quantum jump simulations with the delay function, compared with other numerical methods relying on time discretization (GOBE or time discretized quantum jump simulations). While the computation time for time discretized GOBE typically scales as $\theta^{2.5}$ for one-dimensional σ_+/σ_- VSCPT (see Section I.3.2 in [Bar95]), we observe that, for quantum jump simulations with the delay function, the computation time scales[2] only as $\sim\theta^{0.6}$, *i.e. less rapidly than the interaction time θ* of the simulated experiment (see Section IV.4.4.2 in [Bar95]). This unusual 'acceleration' of the simulation at large times is a numerical manifestation of Lévy statistics. It can be interpreted as the fact that the longer the interaction time, the more probable the occurrence of very long delays, which allows the simulation to reach large interaction times in a small number of steps.

It is therefore possible to generate many individual histories with an interaction time at least equal to θ, and to make ensemble averages at any time smaller than

---

[1] See Appendix A, Section A.1.1.1 for the definitions of Γ and Ω_1.

[2] The precise value of the effective θ exponent might depend on the chosen parameters. If all delays corresponded to trapping times τ_i (no return times $\hat{\tau}_i$), the computation time would be proportional to the number N of trapping events. Thus, it would vary as $\theta^{1/2}$ since $\theta \sim T_N = \sum_{i=1}^{N} \tau_i \propto N^2$. However, the presence of return times $\hat{\tau}_i$, each one composed of many delays (see Section 3.4), adds a contribution (linear in N for the confined recycling model), thereby leading to an effective exponent larger than $1/2$.

θ. One can build in this way the one-dimensional atomic momentum distribution $\pi(p, \theta)$ (see Section 6.2.2), which contains all the information needed for comparison with our statistical approach. We can then extract global quantities characterizing the cooling: the fraction $f_{\text{trap}}(\theta)$ of atoms trapped in quasi-dark states, the half-width $w(\theta)$ and the height $h(\theta)$ of the peak of cooled atoms.

In conclusion, the optimal numerical quantum optics method to study long interaction times, relevant to check the Lévy statistics predictions, is the one using quantum jump simulations. On the other hand, when available, analytical results in the long time asymptotic regime, derived from GOBE, are extremely interesting as benchmarks for our approach.

8.2.3 *Monte Carlo simulations of Raman cooling*

In Raman subrecoil cooling, properly tailored laser pulses are used for implementing a velocity selective excitation scheme. The delay function cannot be easily calculated with an effective Hamiltonian, as is possible in the case of VSCPT which uses a continuous wave laser excitation. One can however solve the time-dependent Schrödinger equation associated with the pulsed laser excitation scheme, and obtain probability of excitation, and thus a jump rate $R(\mathbf{p})$ for an atom with a momentum $\mathbf{p}$. Knowing the momentum change $\Delta\mathbf{p}$ during a jump, one can then perform a Monte Carlo simulation of the atomic momentum evolution [Rei96, RSC01]. This method is numerically efficient, and it has been used to study the role of the shape of the jump rate $R(\mathbf{p})$ around $\mathbf{p} = 0$, in particular of the exponent α of its expansion. Note that this method has also been used to study Raman cooling in a trap [Mor95].

8.3 **Proportion of trapped atoms in one-dimensional** σ_+/σ_- **VSCPT**

We showed in Chapter 5 that the Lévy statistics approach allows one to make precise quantitative predictions of the proportion $f_{\text{trap}}(\theta)$ of trapped atoms, which only depends, in the asymptotic regime of long interaction times, on the tails of the statistical laws describing the trapping times and first return times. It is therefore possible to compare these predictions with the results of the quantum optics methods presented in Section 8.2.2. Since the comparison rests on a single number (more precisely on its evolution with the interaction time), it is particularly easy to make.

The Lévy statistics predictions are fully quantitative, and give not only the laws of variation of $f_{\text{trap}}(\theta)$ with θ, but also the values of the prefactors. The comparison presented in this chapter will therefore be complete, since we know (see Appendix A) the exact correspondence between the parameters such as τ_b and

$\hat{\tau}_b$ characterizing the statistical laws and the physical parameters (laser intensity, atomic parameters) chosen for the quantum jump simulations.

The most complete comparison has been done with the results of the quantum jump simulations of one-dimensional σ_+/σ_- VSCPT, which allow us to check the *Doppler model* (cf. Section 3.2). To check the *unconfined model*, we use the modified quantum jump simulations of one-dimensional σ_+/σ_- VSCPT, where the Doppler effect is made negligible (see Appendix A, Section A.1.1.5, p. 153). A further modification of the quantum jump simulations where the momentum evolution is confined by walls (see Appendix A, Section A.1.2.5, p. 159) allows us to test the *confined model*.

Note that all expressions given below in terms of the atomic parameters correspond to the regime $\Omega_1 \ll \Gamma$ and $\tilde{\delta} = 0$ (see Section A.1.1).

8.3.1 Doppler model

The one-dimensional Doppler recycling model is the one which matches most closely the full one-dimensional σ_+/σ_- VSCPT physics. It corresponds to infinite average trapping times ($\mu = 1/2$) and infinite average first return times ($\hat{\mu} = 1/4$). The prediction of our statistical analysis is given by eq. (5.26):

$$\left(\frac{1}{f_{\text{trap}}(\theta)}\right)_{\text{Lévy}} \xrightarrow{\theta \to \infty} \frac{(\Gamma(3/4))^2}{\Gamma(1/2)} \frac{\hat{\tau}_b^{\hat{\mu}}}{\tau_b^{\mu}} \theta^{1/4}, \tag{8.1}$$

up to a subleading term behaving as $\ln(\theta)$, whose relative contribution becomes negligible at large times.

Using eq. (A.38) for τ_b, eq. (B.23) for $\hat{\tau}_b$ and $\Gamma(1/2) = \sqrt{\pi}$, we obtain

$$\left(\frac{1}{f_{\text{trap}}(\theta)}\right)_{\text{Lévy}} \xrightarrow{\theta \to \infty} \frac{4 \times 0.3296\ldots (\Gamma(3/4))^2}{\pi} \frac{(\Delta p)^{3/2}}{p_0 p_D^{1/2}} \left(\frac{\theta}{\tau_0}\right)^{1/4}. \tag{8.2}$$

Note that the auxiliary parameter p_{trap}, present in both τ_b and $\hat{\tau}_b$, no longer appears in $f_{\text{trap}}(\theta)$, as expected.

It is interesting to interpret the dependences of expression (8.2) on the various parameters. First, τ_0 appears as it should as the only time to which the total time θ compares. Second, $f_{\text{trap}}(\theta)$ depends on three momentum parameters, Δp, p_0 and p_D. The rms length Δp of the elementary step in the p-space controls the number of steps needed for a successful return to the trap (see Section 3.4.2). If Δp is larger, an atom coming back to the vicinity of $p = 0$ misses the trap more often and therefore f_{trap} is expected to be smaller, which agrees with eq. (8.2). As for the momentum p_0, it is related to the width of the dark resonance dip around $p = 0$. More precisely, for larger p_0, $R(p \simeq 0) = (p/p_0)^2/\tau_0$ (cf. eq. (3.5)) is smaller and the time spent in the trap is thus larger. Therefore the number of trapped atoms $f_{\text{trap}}(\theta)$ is expected to be larger, which again agrees with eq. (8.2). Finally, the Doppler scale p_D gives the momentum scale for the decay of $R(p)$ at large p. If p_D is larger, the decay of $R(p)$ is slower, the return times $\hat{\tau}$ are thus smaller and therefore $f_{\text{trap}}(\theta)$ is expected to be larger, which again agrees with eq. (8.2).

For numerical purposes, $f_{\text{trap}}(\theta)$ is more conveniently expressed using eq. (A.39) for τ_b, and eq. (A.44) for $\hat{\tau}_b$. Using $\Gamma(3/4) = 1.225\,41\ldots$, we get

$$
\left(\frac{1}{f_{\text{trap}}(\theta)}\right)_{\text{Lévy}} \xrightarrow{\theta\to\infty} \frac{0.3296\ldots(\Gamma(3/4))^2 2^{27/4}}{\pi\, 3^{3/4}} \left(\frac{\Omega_R}{\Omega_1}\right)^{3/2} (\theta\Gamma)^{1/4}
$$

$$
= 7.44\ldots \left(\frac{\Omega_R}{\Omega_1}\right)^{3/2} (\theta\Gamma)^{1/4}. \tag{8.3}
$$

We can now compare this result with other approaches.

First, our statistical result is in agreement[3] with the asymptotic analytical solution of [AlK96]. Such an agreement is remarkable and far from trivial. Indeed the jump rate of our simple Doppler model is fairly different from the real jump rate of quantum optics (see Appendix A). We believe that the reason for this excellent agreement is that both jump rates coincide around $p = 0$ and for $p \to \infty$, i.e. precisely *in the regions determining the broad distributions* characterizing respectively the long trapping times and the long recycling times, which are both relevant quantities for determining f_{trap}. The merit of our very simplified statistical models is to retain the features relevant to the asymptotic regime.

Note also that the analytical result obtained in [SSY97] using kinetic equations derived from GOBE gives the correct asymptotic $\theta^{-1/4}$ dependence for f_{trap}. The numerical prefactor, however, is 40% larger than expression (8.3), thus in clear disagreement with our statistical result, but also with the analytical result of [AlK96] and with the numerical simulations described below.

Second, we compare our statistical result (8.3) with numerical calculations using quantum jump simulations[4]. Equation (8.3) suggests plotting $1/f_{\text{trap}}$ as a function of $\theta^{1/4}$: the data should fall on a straight line for large θ, up to a slowly varying logarithmic correction. The numerical results are shown in Fig. 8.1 for a laser intensity corresponding to a Rabi frequency $\Omega_1 = 0.3\Gamma$ and a recoil frequency $\Omega_R = \Gamma/37.5$ (detuning $\delta = 0$). Compared with our previously published data [BBE94, Bar95], these results give access to interaction times two orders of magnitude larger, further into the asymptotic regime. After a transient regime for $(\Gamma\theta)^{1/4} \leq 10$, the simulated $1/f_{\text{trap}}(\theta)$ (black dots in Fig. 8.1) is in remarkable

---

[3] The residual discrepancy between our statistical approach and eq. (28) of [AlK96] is about 0.5%, which is less than the uncertainty of 3% on eq. (28) of [AlK96] (which results from an approximate integral estimation).

[4] As indicated in Section 8.2.2, numerical calculations based on Generalized Optical Bloch Equations are too demanding in computer power to reach the long time asymptotic regime, and have thus not been used here.

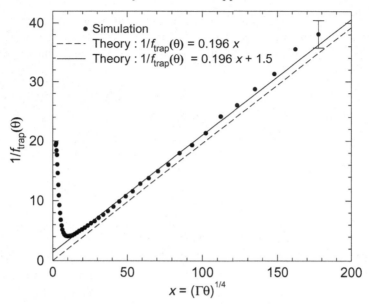

Fig. 8.1. Doppler case. Proportion of trapped atoms, obtained by a quantum jump simulation, for the one-dimensional σ_+/σ_- VSCPT scheme (case of metastable helium, with $\Omega_R/\Gamma = 1/37.5$, $\Omega_1/\Gamma = 0.3$, $\delta = 0$, initial momentum distribution width $\delta p_{rms} = 2\hbar k$, $p_{trap} = 0.08\hbar k$, number of samples $N_{samp} = 4 \times 10^4$). We have plotted directly $1/f_{trap}(\theta)$ as a function of $(\Gamma\theta)^{1/4}$, as suggested by eq. (8.3). We have shown for comparison the best linear fit (for $x \geq 15$) with the theoretical slope (0.196) (solid line). This assumes that the subleading logarithmic term can be replaced by a constant, found to be 1.5 (see text). The dashed straight line corresponds to eq. (8.3) without any subleading terms. The error bar on the last point of the simulation (which is the largest) corresponds to $\pm 2\sigma_{1/f_{trap}}$.

agreement with the predicted asymptotic behaviour of $(1/f_{trap}(\theta))_{\text{Lévy}}$ given by eq. (8.3) (dashed line). The agreement is improved if one allows for the subleading (logarithmic) term, which we approximate by a constant in the investigated time interval. This leads to the solid line shown in Fig. 8.1. The numerical points slightly depart from this line at long times as expected, since this reflects the difference between a constant subleading term and the actual logarithmic behaviour.

Note also that for small f_{trap}, the statistical uncertainties are amplified in this representation since $\sigma_{1/f_{trap}} = \sigma_{f_{trap}}/f_{trap}^2 \simeq 1/(f_{trap}^{3/2} N_{samp}^{1/2})$. Indeed, the standard deviation $\sigma_{f_{trap}}$ for f_{trap} (which follows a Bernoulli process) is given at large N_{samp} and small f_{trap} by the standard deviation of the binomial law:

$$\sigma_{f_{trap}} = \sqrt{\frac{f_{trap}(1 - f_{trap})}{N_{samp}}} \simeq \sqrt{\frac{f_{trap}}{N_{samp}}}, \qquad (8.4)$$

where N_{samp} is the number of atoms used in the simulation.

Similar quantum jump simulations by Fioretti *et al.* [FZA95] have explored several properties of one-dimensional σ_+/σ_- VSCPT. In particular, these authors have shown that if one changes the width of the initial momentum distribution, the first part of the evolution as well as the maximum value of $f_{\text{trap}}(\theta)$ are different but they tend to converge towards the same asymptotic decay. We have also checked that the results of the simulations displayed in [FZA95] are in reasonable agreement (within 20% at $\theta = 10^8 \Gamma^{-1}$) with the asymptotic prediction of our Lévy statistics analysis (eq. (8.3) above).

Finally, GOBE numerical solutions have been used to confirm qualitatively the decay of $f_{\text{trap}}(\theta)$ at large θ, both on the same atomic transition, $J = 1 \to J = 1$, that we used in our numerical studies, and on a $J = 2 \to J = 2$ transition [MZL96].

8.3.2 Unconfined model

For the unconfined model in which the Doppler effect is not included, our statistical theory predicts (cf. eq. (5.27))

$$\left[f_{\text{trap}}(\theta) \right]_{\text{Lévy}} \xrightarrow[\theta \to \infty]{} \frac{\tau_b^{1/2}}{\tau_b^{1/2} + \hat{\tau}_b^{1/2}}. \tag{8.5}$$

Using eq. (A.38) for τ_b and eq. (A.46) for $\hat{\tau}_b$, we obtain

$$\left[f_{\text{trap}}(\theta) \right]_{\text{Lévy}} \xrightarrow[\theta \to \infty]{} \left(1 + \frac{4}{\pi \sqrt{2}} \frac{\Delta p}{p_0} \right)^{-1}. \tag{8.6}$$

This expression is remarkably simple. Whereas the theory has been developed for *time* variables in previous chapters, eq. (8.6) expresses the trapped population in terms of two *geometric* parameters of the random walk, the size p_0 of the dip of $R(p)$ and the rms length Δp of an elementary step. Note that one can interpret the dependence of eq. (8.6) on these parameters as in Section 8.3.1.

For numerical purposes, f_{trap} is more conveniently expressed using eq. (A.39) for τ_b and eq. (A.47) for $\hat{\tau}_b$ to get

$$\left[f_{\text{trap}}(\theta) \right]_{\text{Lévy}} \xrightarrow[\theta \to \infty]{} \left(1 + \frac{32}{\pi \sqrt{3}} \frac{E_R}{\hbar \Gamma} \frac{\Gamma^2}{\Omega_1^2} \right)^{-1}. \tag{8.7}$$

With the same parameters as in Section 8.3.1, we obtain numerically

$$\left[f_{\text{trap}}(\theta \to \infty) \right]_{\text{Lévy}} \simeq 0.365 \pm 0.02. \tag{8.8}$$

The uncertainty comes from the approximations made in the diagonalization of the Hamiltonian which affect τ_b and $\hat{\tau}_b$. It could be reduced if necessary.

The results of the quantum jump simulations are plotted in Fig. 8.2. In these simulations, the Doppler effect has been made negligible following the method

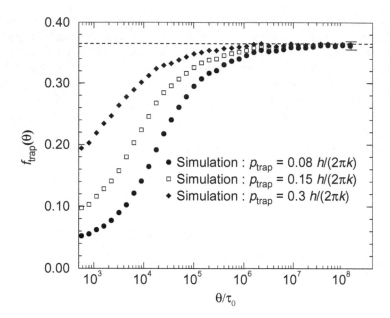

Fig. 8.2. *Unconfined case.* Proportion of trapped atoms, obtained by a quantum jump simulation, for one-dimensional σ_+/σ_- VSCPT, with the same parameters as in Fig. 8.1, but with the Doppler effect made negligible (see explanation in Section A.1.1.5, p. 153). The proportion of trapped atoms tends towards the constant f predicted by the Lévy statistics analysis (horizontal line). The various curves correspond to different choices of the arbitrary parameter p_{trap} and the asymptotic value is clearly independent of this parameter. The error bar on the last point corresponds to $\pm 2\sigma_{f_{trap}}$ for a finite number of atoms $N_{samp} = 2 \times 10^4$ (see eq. (8.4)).

explained in Section A.1.1.5, p. 153. Taking $F = 316$ (see eq. (A.19a,b)), the scale at which the jump rate $R(p)$ is constant is of the order of $3000\hbar k$, larger than the typical maximum momentum $\simeq 1500\hbar k$ for the atoms diffusing out of the trap. Figure 8.2 shows the evolution of the fraction of trapped atoms, for several values of the parameter p_{trap} characterizing the arbitrarily defined trapping region. The proportion of trapped atoms clearly tends towards a constant value, independent of p_{trap}

$$\left[f_{trap}(\theta \to \infty) \right]_{sim} = 0.363 \pm 0.003, \tag{8.9}$$

where the indicated uncertainty is the standard deviation (one σ) of the statistical fluctuation of the 2×10^4 simulations. This result is in excellent agreement with the prediction (eq. (8.8)) of our statistical theory.

The analytic solution of GOBE presented in [AlK92, AlK93] is relevant to the unconfined model[5]. These authors found that f_{trap} tends towards a constant, explicitly expressed as a function of the atom and laser parameters. One finds that their result is identical to eq. (8.7) with $\tilde{\delta} = 0$. The agreement is therefore perfect.

8.3.3 Confined model

The one-dimensional confined model is adequate for both one-dimensional VSCPT with friction and one-dimensional Raman cooling with friction. For VSCPT ($\alpha = 2$), it corresponds to an infinite average trapping time ($\mu = 1/2$) and a finite average first return time. Our statistical analysis then predicts (cf. eq. (5.23))

$$f_{\text{trap}}(\theta) \underset{\theta \to \infty}{=} 1 - \frac{1}{\pi} \frac{\langle \hat{\tau} \rangle}{\tau_b^{1/2} \, \theta^{1/2}} + \cdots. \tag{8.10}$$

Thus, all atoms are expected to be trapped at long times, with the untrapped proportion decaying as $1/\theta^{1/2}$.

Using eq. (A.38) for τ_b and eq. (3.56) for $\langle \hat{\tau} \rangle$, we get

$$f_{\text{trap}}(\theta) \underset{\theta \to \infty}{=} 1 - \frac{4}{\pi^{3/2}} \frac{p_{\max}}{p_0} \left(\frac{\tau_0}{\theta} \right)^{1/2} + \cdots. \tag{8.11}$$

Note that the auxiliary parameter p_{trap}, present in both τ_b and $\langle \hat{\tau} \rangle$, no longer appears in $f_{\text{trap}}(\theta)$, as expected.

The amplitude of the $1/\theta^{1/2}$ term describing the filling of the trap in eq. (8.11) can be interpreted physically. As above, the time τ_0 between jumps outside the trapping region appears as a scaling time, and p_0 characterizes the width of the trapping dip. Since $p_{\max}$ gives the position of the walls in momentum space, the atoms return more rapidly to the trapping region when $p_{\max}$ is smaller, and the trap is thus expected to fill up more rapidly. This is in agreement with eq. (8.11). Finally, the absence of the length Δp of the random walk step has already been discussed in Section 3.4.4.

For numerical purposes, we replace in eq. (8.10) τ_b by eq. (A.39), and $\langle \hat{\tau} \rangle$ by eq. (3.56) (with eq. (A.23) for τ_0). We obtain

$$f_{\text{trap}}(\theta) \underset{\theta \to \infty}{=} 1 - \frac{32}{\pi^{3/2}} \frac{p_{\max}}{\hbar k} \frac{\Omega_R \Gamma^2}{\Omega_1^3} \frac{1}{(\Gamma \theta)^{1/2}} + \cdots. \tag{8.12}$$

---

[5] Further GOBE analytic results [AIK96] pointed out an heuristically expected connection between the Doppler model and the unconfined model. In the Doppler model (with a not too broad initial momentum distribution), $f_{\text{trap}}(\theta)$ first increases during a limited time, *then reaches a maximum given by the asymptotic solution (8.7) of the unconfined model*, and finally decreases following eq. (8.3) [AIK96]. Indeed, as explained in Section A.1.1.5 (p. 153), the unconfined model and the Doppler model are equivalent at intermediate times because the atoms diffusing out of the trap have not yet reached large enough p values to feel the Doppler decrease of $R(p)$. This property is used in our quantum jump calculation to simulate the unconfined model by a Doppler model (full quantum problem) in which the Doppler decay of $R(p)$ is pushed to large enough values of p.

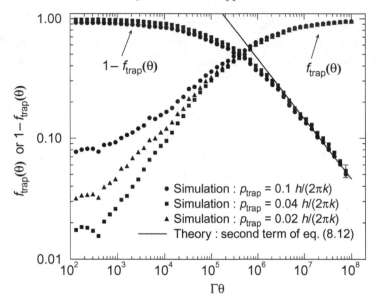

Fig. 8.3. *Confined model.* Proportion of trapped atoms, obtained by a quantum jump simulation, for one-dimensional σ_+/σ_- VSCPT, ($\Omega_R = \Gamma/37.5$, $\Omega_1 = 0.1\Gamma$, $\delta = 0$), with walls confining the momentum diffusion in the interval $[-3\hbar k, 3\hbar k]$. Each curve corresponds to different choices of the arbitrary parameter p_{trap}: the asymptotic value is clearly independent of this parameter. The set of increasing curves gives $f_{trap}(\theta)$, which clearly tends to one, as predicted by our statistical analysis. The set of decreasing curves corresponds to $1 - f_{trap}(\theta)$ (note the log–log scale). They asymptotically match the predicted $\theta^{-1/2}$ evolution indicated by the straight line (no adjustable parameter). The error bar corresponds to $\pm 2\sigma_{f_{trap}}$, with a finite number of atoms $N_{samp} = 4942$ (cf. eq. (8.4)).

This result can be compared with quantum jump simulations with walls at $\pm p_{max}$ (see Section A.1.2.5, p. 159) and with the Doppler effect made negligible. Figure 8.3 displays $f_{trap}(\theta)$ in such a simulation with $p_{max} = 3\hbar k$, $\Omega_R = \Gamma/37.5$, $\Omega_1 = 0.1\Gamma$, $\delta = 0$ and various p_{trap} values. First, it shows that $f_{trap}(\theta)$ indeed tends to one, as the first term of eq. (8.12) indicates. Then, the second term of eq. (8.12), giving the filling behaviour of the trap, also perfectly matches the results of the simulations. This confirms both the $1/\theta^{1/2}$ power-law behaviour and the prefactor of this power law with an accuracy of about one per cent.

Other quantum jump simulations [SHP93, WEO94, MDT94] have considered VSCPT schemes with friction, treating the problem fully quantum mechanically. This is more rigorous than our introduction of walls in momentum space to model friction. These simulations confirm qualitatively that the proportion of trapped atoms is larger with friction than without it. However, they do not reach long enough times to show that the trapped proportion tends to one ($f_{\text{trap}}(\theta) < 0.8$ in these simulations), or to obtain the asymptotic filling behaviour of the trapped proportion (second term of eq. (8.12)).

8.4 Width and shape of the peak of cooled atoms

8.4.1 Statistical predictions

We have shown in Chapter 6 how the statistical analysis allows one to predict the explicit form of the momentum distribution of the cooled atoms. We have in particular considered the very favourable case (in the perspective of efficient cooling), where the trapping times distribution is broad, while friction makes the recycling times finite. In this case we have found that the momentum distribution can be deduced from a universal function by scaling laws only depending on θ, α and μ (Section 6.3). In particular, the momentum distribution is predicted to have tails broader than a Maxwellian distribution, scaling as $p^{-\alpha}$ (i.e. as p^{-2} for VSCPT); the half-width $w(\theta)$ is expected to vary as $\theta^{-1/\alpha}$ (i.e. as $\theta^{-1/2}$ in the case of VSCPT) independently of the dimensionality. Finally, the momentum distribution is predicted to be flatter near its centre than the normalized Lorentzian with the same tails. These predictions have been subjected to various tests that we now discuss.

8.4.2 Comparison to quantum calculations

We have performed quantum jump simulations in the case of one-dimensional σ_+/σ_- VSCPT with friction simulated by walls restricting the momentum evolution to the interval $[-p_{\max}, +p_{\max}]$ (Section A.1.2.5, p. 159). The corresponding statistical model is the one-dimensional *confined model*, with $\alpha = 2$ (and therefore $\mu = 1/2$). The simulation has been done with the same parameters as for Fig. 8.3, and Fig. 8.4 shows the momentum profile at $\theta = 2 \times 10^8 \Gamma^{-1}$, definitely in the asymptotic regime of long interaction times.

The analytical prediction of the statistical analysis (eq. (6.31) and eq. (6.32)) is obtained by substituting eq. (3.34b) for $\mathcal{A}_\mu$ (with $\Gamma(1/2) = \sqrt{\pi}$), eq. (3.25) for C_D, eq. (6.2) for p_θ and eq. (A.26) for $\tau_0 p_0^2$:

$$\pi(p, \theta) = \frac{8}{\pi^{3/2}\hbar k} \frac{\Omega_R}{\Omega_1} \sqrt{\Gamma\theta} \, \mathcal{G}\left(4\frac{\Omega_R}{\Omega_1} \sqrt{\Gamma\theta} \frac{p}{\hbar k}\right). \tag{8.13}$$

This prediction is plotted as the solid curve in Fig. 8.4.

The agreement between the quantum jump simulation and the statistical prediction, which involves no adjustable parameter, is remarkable. The profile predicted by the confined statistical model is clearly discriminated against the pure Gaussian or Lorentzian profiles that fit the results of the simulation less well. In particular,

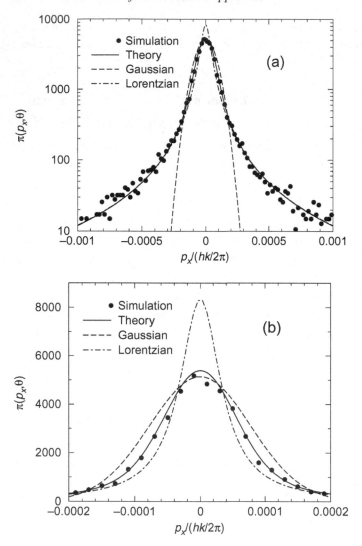

Fig. 8.4. Momentum profile obtained by a quantum jump simulation with confining walls at $\pm p_{max} = 3\hbar k$. The interaction time is $\theta = 2 \times 10^8 \Gamma^{-1}$. The parameters are identical to those of Fig. 8.3 ($\Omega_R/\Gamma = 1/37.5$, $\Omega_1/\Gamma = 0.1$, $\delta = 0$). Number of samples: $N_{samp} = 23\,340$. (a) Large scale figure revealing the tails of the peak (note the logarithmic scale for $\pi(p_x, \theta)$). (b) Small scale figure revealing the central part of the peak (note the linear scale for $\pi(p_x, \theta)$). The profile predicted by the statistical model (solid curve) fits the result of the simulation better than a normalized Gaussian $y = \exp(-x^2/(2\sigma_G^2))/(\sqrt{2\pi}\sigma_G)$ adjusted on the centre (in the interval $[-2 \times 10^{-4}\hbar k, 2 \times 10^{-4}\hbar k]$), or a normalized Lorentzian $y = \left(\pi\sigma_L \left(1 + (x/\sigma_L)^2\right)\right)^{-1}$ adjusted on the tails (in the intervals $[-10^{-3}\hbar k, -3 \times 10^{-4}\hbar k]$ and $[3 \times 10^{-4}\hbar k, 10^{-3}\hbar k]$). Note the flat top of the profile, as compared with the normalized Lorentzian adjusted on the tails.

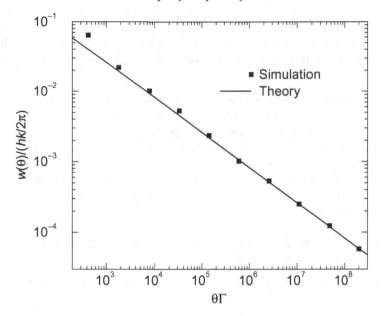

Fig. 8.5. Momentum distribution half-width $w(\theta)$ as a function of the interaction time θ, in the case of one-dimensional σ_+/σ_- VSCPT with confining walls. Same parameters as in Fig. 8.4. The squares are the results of the quantum jump simulations ($N_{\text{samp}} = 23\,340$), where the width is obtained by fitting the result of the simulation to the expected profile (the uncertainty on the value of $w(\theta)$ is $\simeq 10$–20%). The solid line is the statistical prediction of a $\theta^{-1/2}$ dependence, with a prefactor determined without any adjustable parameter.

the flattening around $p = 0$, which is a specific prediction of the statistical approach, is clearly observed on the quantum jump simulations. Note also that the tails decay as p^{-2}, i.e. as a Lorentzian, in agreement with eq. (6.36).

In order to make a complementary test of the scaling laws with θ, we have extracted from the quantum jump simulations the half-width at $e^{-1/2}$ of the profiles $w(\theta)$ at various interaction times θ (see Fig. 8.5). The prediction of the statistical analysis, indicated by the solid line, is obtained using eq. (6.33), together with eq. (6.2) and eq. (A.26) to give:

$$w(\theta) = 0.22\ldots \frac{\Omega_1}{\Omega_R} \frac{1}{(\theta\Gamma)^{1/2}}\, \hbar k. \tag{8.14}$$

The agreement between the statistical analysis and the simulations is within the statistical accuracy of the simulation, without any adjustable parameter. This result supports in particular the prediction of a $\theta^{-1/2}$ scaling of the width of the peak of cooled atoms.

These simulations might also allow us to test the variation of other quantities, for example the height of the peak as a function of θ. However, such tests would be redundant, since we have already checked that the profile and the width follow the predicted law. The presented results therefore allow us to conclude that full agreement exists between the quantum jump simulations and the statistical predictions.

Although they do not allow one to reach very long interaction times, numerical integrations of the GOBE have also shown the expected $\theta^{-1/2}$ evolution of $w(\theta)$ in the time range explored. Results have been obtained in one-dimensional σ_+/σ_- VSCPT [AAK89], and in two-dimensional VSCPT [MaA91, MaA92]. Importantly, these simulations also agree with our results on the evolution of the height $h(\theta)$ of the cooled peak (see table 6.1): $h(\theta) \simeq \theta^{1/2}$ in one dimension [AAK89] and $h(\theta) \simeq \theta$ in three dimensions [MaA91] (the variation of the subleading term in $\log \theta$ being negligible at the relatively short time scales that were investigated).

More significant for the asymptotic regime, the analytical results of [AlK93], applicable to the unconfined model, give the same shape $\mathcal{G}(p)$ for the momentum distribution and the same scaling laws for the width $w(\theta)$ and the height $h(\theta)$ as those obtained in Section 6.5.

8.4.3 Experimental tests

Several experimental results on VSCPT in one, two and three dimensions, agree with the predictions of the statistical analysis, within the limited experimentally accessible range of interaction times. Before analysing the results, we clarify which recycling models are tested by these experiments. In one dimension, all the results presented in this section have been performed in the σ_+/σ_- configuration, which corresponds to the Doppler model at very long times. However, on the relatively short time scale explored experimentally, the atomic momenta remain small enough for the Doppler effect to be negligible and thus these experiments test the one-dimensional *unconfined model* (see footnote [5] in Section 8.3.2 and Section A.1.1.5, p. 153). In two and three dimensions, the experiments have been performed in configurations creating friction to obtain a significant number of cooled atoms (see Section 8.5); in this case the *confined model* is relevant.

The *width* of the cooled peak has been studied in one, two and three dimensions [AAK88, BSL94, LKS95]. The predicted $\theta^{-1/2}$ behaviour was observed in all three cases. A new method for measuring ultranarrow momentum distributions, based on the measurement of the spatial correlation function of the cold atoms [SHK97], has been used to check experimentally the validity of the $\theta^{-1/2}$ law for one-dimensional σ_+/σ_- VSCPT (Fig. 8.6), up to interaction times of $\theta \simeq 2 \times 10^4 \Gamma^{-1}$; for the longest interaction time, the corresponding temperature of the cooled atoms was $T \simeq T_R/800$.

In order to compare the theory with experiments, eq. (8.14) must be transformed

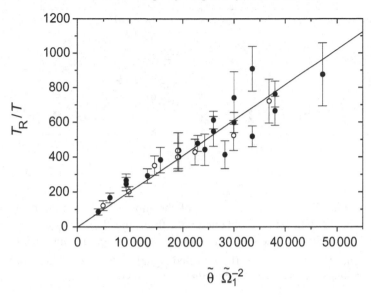

Fig. 8.6. Experimental variation of the inverse of the effective temperature T with the parameter $\tilde{\theta}\,\tilde{\Omega}_1^{-2}$ (notation: $\tilde{\theta} = \theta\Gamma$, $\tilde{\Omega}_1 = \Omega_1/\Gamma$; parameters: $\Omega_R/\Gamma = 1/37.5$, $\delta = 0$). The filled circles correspond to $\tilde{\Omega}_1 = 0.72(2)$ and $\tilde{\theta}$ varying from 2000 to 25 000. The open circles correspond to $\tilde{\theta} = 10^4$ and $\tilde{\Omega}_1$ varying from 0.5 to 2. The line is the theoretical prediction of the statistical analysis, without any adjustable parameter, eq. (8.15). (Figure reproduced from figure V.11 in [Sau98].)

into:

$$\frac{T_R}{T} = B\left(\frac{\hbar k}{w(\theta)}\right)^2 = \frac{B}{(0.22\ldots)^2}\left(\frac{\Omega_R}{\Gamma}\right)^2\frac{\tilde{\theta}}{\tilde{\Omega}_1^2}, \qquad (8.15)$$

where $B \simeq 1.45$ is a numerical factor taking into account the fact that the experimental temperature T was inferred from the experimental data by a fit of the momentum distribution to a Lorentzian instead of the accurate expression eq. (8.13) (see Sections V.4.2 and VI.4.5 in [Sau98]). The parameters $\tilde{\Omega}_1$ and $\tilde{\theta}$ are defined in the caption of figure 8.6. As can be seen in Fig. 8.6, the prefactor of the theoretical $\theta^{-1/2}$ law agrees very well with the experimental results.

The *shape* of the momentum distribution has been studied in one-dimensional σ_+/σ_- VSCPT. Experiments first showed that it is clearly different from a Maxwellian, and that the tails are well fitted by a Lorentzian [Bar95]. Recently, the new method of [SHK97] has allowed a more precise experimental study of the shape, and important results have been obtained [Sau98, SLC99]. The quantity $\Pi_{NC}(t_D)$ that is actually measured is related to the Fourier transform of the momentum

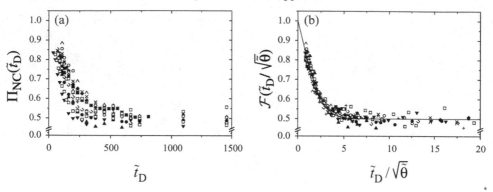

Fig. 8.7. Experimental test of the scaling of the momentum distribution with θ (from [SLC99]). The 12 data sets are obtained with the same Rabi frequency $\tilde{\Omega} = 0.72(2)$ but with different values of the interaction time $\tilde{\theta} = \theta\Gamma$ ranging from 2000 to 15 000. (a) Measured signal $\Pi_{NC}(\tilde{t}_D = t_D\Gamma)$. (b) Rescaled signal $\mathcal{F}(\tilde{t}_D/\sqrt{\tilde{\theta}})$. After rescaling, all the data clearly collapse to the theoretical prediction (solid curve).

distribution $\pi(p, \theta)$:

$$\Pi_{NC}(t_D) = \frac{1}{2} + \frac{1}{2}\int_{-\infty}^{+\infty} dp\, \pi(p, \theta) \cos\left(\frac{2kp}{M}t_D\right), \qquad (8.16)$$

where t_D is the parameter of the measurement (see [SLC99] for details). Using the expression for $\pi(p, \theta)$ obtained in Chapter 6 (see case $\mu = \hat{\mu}$ in Section 6.5), and writing $u = p/p_\theta$, we first obtain

$$\Pi_{NC}(t_D) = \frac{1}{2} + \frac{1}{2}p_\theta h(\theta) \int_{-\infty}^{+\infty} du\, \mathcal{G}(u)\, \cos\left(\frac{2kp_\theta u}{M}t_D\right). \qquad (8.17)$$

In one dimension, with $\alpha = 2$ and the unconfined model, the quantity $p_\theta h(\theta)$ does not depend on θ: $p_\theta h(\theta) = C$. Moreover, using eq. (6.2) for p_θ and eq. (A.25) for $p_0\sqrt{\tau_0}$, we have $2kp_\theta t_D/M = \tilde{\Omega}_1\tilde{t}_D/\sqrt{\tilde{\theta}}$ with $\tilde{\Omega}_1 = \Omega_1/\Gamma$, $\tilde{t}_D = t_D\Gamma$ and $\tilde{\theta} = \theta\Gamma$. Thus the measured quantity $\Pi_{NC}(t_D)$ depends only on θ and t_D through $\tilde{t}_D/\sqrt{\tilde{\theta}}$ and obeys the scaling relation:

$$\Pi_{NC}(t_D) = \mathcal{F}\left(\frac{\tilde{t}_D}{\sqrt{\tilde{\theta}}}\right) = \frac{1}{2} + \frac{1}{2}C\int_{-\infty}^{+\infty} du\, \mathcal{G}(u)\cos\left(\frac{\tilde{t}_D}{\sqrt{\tilde{\theta}}}\tilde{\Omega}_1 u\right). \qquad (8.18)$$

The experimental signals should therefore be self-similar and reducible to the universal distribution predicted by the statistical analysis of Chapter 6. The experimental results presented in Fig. 8.7 clearly confirm this important point.

Moreover, with the method of [SHK97], the *shape itself* of the momentum distribution can also be studied precisely enough to be compared with the shape predicted in Section 6.3. Figure 8.8 shows that the predicted shape of $\mathcal{F}$ agrees

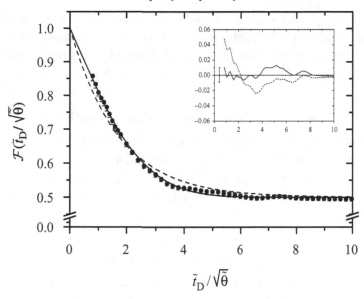

Fig. 8.8. Fit of the averaged data with the calculated function $\mathcal{F}$ (solid curve) or a simple exponential for $\mathcal{F} - 1/2$, corresponding to a Lorentzian momentum distribution (dashed curve). The data points are obtained after averaging the rescaled curves of Fig. 8.7b. The fit with the theoretical prediction is clearly better, as shown by the residues presented in the inset. The vertical bar on the left is the largest error bar of the data. The data are well adjusted by the theoretical function ($\chi^2 = 0.0049$) and the fit gives $\widetilde{\Omega} = 0.70$ in very good agreement with the measured value, $\widetilde{\Omega} = 0.72(2)$. The exponential fit leads to a much poorer $\chi^2 = 0.0285$.

very well with the experimental results. Note that $\mathcal{F} - 1/2$ is the Fourier transform of the momentum distribution. If this momentum distribution was a Lorentzian, one would get an exponential for $\mathcal{F} - 1/2$. The best exponential is represented as a dotted curve in Fig. 8.8. One can clearly see that the experimental results are in much better agreement with the Fourier transform of the theoretical shape computed in Chapter 6. This confirms the flattening of $\pi(p, \theta)$ for $p \simeq 0$ compared with a Lorentzian.

To sum up, for one-dimensional σ_+/σ_- VSCPT, the comparison between the experiments and the statistical description is complete and satisfactory. Note however that the fraction of cooled atoms (i.e. the absolute area of the momentum distribution) has not yet been studied experimentally.

Significant experimental results have also been obtained in the case of Raman subrecoil cooling [DLK94, LAK96]. Remarkably, it has been possible to control experimentally the exponent α of the jump rate around $p = 0$, by a proper choice of pulse sequences [RBB95]. In this experiment, it has been possible to compare the cases $\alpha = 2$ and $\alpha = 4$, in a one-dimensional configuration. The measured width $w(\theta)$ clearly behaves as $\theta^{-1/2}$ and $\theta^{-1/4}$ respectively, as expected from the Lévy statistics analysis of Chapter 6 (table 6.1). The statistical predictions have also been found to be in agreement with the Monte Carlo simulations of Raman cooling [Rei96, RSC01].

8.5 Role of friction and of dimensionality

A very important result of Chapters 5 and 6 is that the magnitude of the peak of cooled atoms (in contrast to its width), is dramatically influenced by the existence of friction forces and by the dimensionality of the problem. We gather here the results concerning the roles of friction and dimensionality that provide a test of these predictions.

8.5.1 One-dimensional case

According to the statistical analysis of Chapter 5, the role of friction is not crucial in one-dimensional VSCPT. More precisely, at times short enough that atoms outside the trap have not reached too large values of p, the three different models of recycling – Doppler, unconfined, confined – do not give dramatically different predictions for the proportion of trapped atoms. This prediction is clearly supported by our quantum jump simulations of one-dimensional σ_+/σ_- VSCPT (cf. Section 8.3) as shown in Fig. 8.9: it is only for interaction times larger than $10^4 \Gamma^{-1}$ (this is about the longest time range experimentally investigated to date) that the simulations give results differing by more than a factor of three.

This prediction of a non-crucial role of friction in one dimension is also supported by other quantum jump simulations [MDT94], which consider a situation where friction is fully included in the quantum treatment of the problem. These simulations consider a one-dimensional lin/45°/lin VSCPT situation, where counterpropagating waves have linear polarizations making a 45° angle and create a friction force. They show that the friction helps somewhat, but does not play a dramatic role within the investigated time range.

Experimentally, the limited role of friction in one-dimensional subrecoil cooling is confirmed by the fact that efficient subrecoil cooling has been observed in one-dimensional σ_+/σ_- VSCPT, where there is no friction [AAK88, BSL94, KSP97], while an experiment of one-dimensional VSCPT cooling with friction [ESW96] does not show a dramatic improvement in the cooling efficiency.

8.5.2 Higher dimensional case

According to our Lévy statistics analysis, friction is expected to play a crucial role in two- and three-dimensional VSCPT ($\alpha = 2$, i.e. quadratic jump rate around

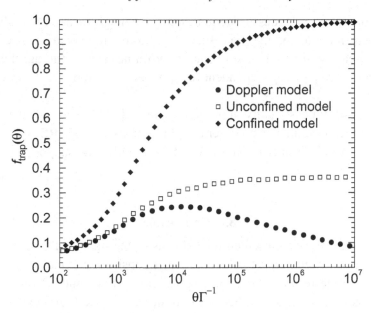

Fig. 8.9. *Influence of the recycling in one dimension.* All simulations have been carried out with $\delta = 0$, $\Omega_1/\Gamma = 0.3$, $\Omega_R/\Gamma = 1/37.5$, initial momentum distribution width $\delta p_{\rm rms} = 2\hbar k$; $p_{\rm trap} = 0.08\hbar k$. Number of samples $N_{\rm samp} = 4 \times 10^4$ for the Doppler model, 2×10^4 for the unconfined model and 4×10^4 for the confined model (with $p_{\rm max} = 3\hbar k$). The simulations corresponding to the three models are detailed in Section 8.3.

$p = 0$). As soon as D is larger than or equal to two, we have $\mu \geq 1$ so that the average trapping time does not diverge. The fraction of trapped atoms will therefore be significant only when the average recycling time is also finite, i.e. in the case where there is a friction. More precisely, we have shown in Chapter 5 and Appendix C[6] that, in the presence of friction, the fraction of trapped atoms should slowly tend to one in two dimensions and to a certain constant less than one in three dimensions. On the other hand, without friction, the fraction of trapped atoms should rapidly tend to zero, faster than $1/\theta$.

This statistical prediction is qualitatively supported by the quantum optics calculation of [MaA91], a rare example of a two-dimensional calculation based on GOBE, which suggests the importance of friction, as pointed out by the authors.

Experiments yield very convincing evidence of the predicted crucial role of friction in two- or three-dimensional VSCPT [LBS94, LKS95, KSP97]. In these experiments, the sign of the p-dependent force (Sisyphus force) at moderate values of p is controlled by the sign of the laser detuning. Significant subrecoil cooling could be observed experimentally only in the case of a positive detuning, where

---

[6] See also Section V.4.5 in [Bar95].

the Sisyphus force provides a friction effect[7]. With zero detuning, there is no
p-dependent force to confine the diffusion of atomic momenta and, accordingly,
no cooled peak is observed experimentally. With negative detuning, the situation
is even worse since then p-dependent force expels the atoms from the vicinity of
$p = 0$.

The roles of dimensionality and friction have also been tested in the case of
Raman subrecoil cooling, by numerical simulations [Rei96, RSC01], based on
the method explained in Section 8.2.3. These simulations support the statistical
predictions.

8.6 Conclusion

In this chapter, we have presented several tests of the statistical analysis of subrecoil cooling presented in this book. These tests consist of comparing the statistical
predictions to the results of quantum optics calculations and/or to experimental
results. Whenever comparison has been possible, we have found excellent agreement, in some cases of the order of one per cent, not only for the exponent of
the scaling laws, but also for the prefactors: in other words, agreement is reached
without any adjustable multiplicative factor.

In particular, we have found clear confirmation of the $\theta^{-1/2}$ dependence of
the width of the momentum distribution of cooled atoms, i.e. of the indefinite
decrease of the temperature with time. The efficiency of the cooling process,
characterized by the fraction of trapped atoms, has also been found in agreement
with the statistical predictions for the three recycling models. We have also tested
in detail another remarkable and specific prediction of the statistical analysis: in the
asymptotic regime of long interaction times, the atomic momentum distribution
obeys a well defined scaling law; its shape can be described as a Lorentzian
profile flattened around its maximum. As discussed in Chapter 7, these features
reflect the non-ergodicity of the cooling process studied here, and it is remarkable
that these properties have been confirmed both by experiments and by quantum
simulations. Finally, the crucial role of the dimensionality and of the presence of
friction forces (an important prediction of the statistical approach) is confirmed by
existing experimental and numerical results.

It is remarkable that some of our results are found to be asymptotically *exact*
in spite of the relatively unsophisticated description of the momentum random
walk (cf. Section 3.2). This brings an interesting explanation in the context of

---

[7] Notice, however, that the real situation significantly departs from the simplified model. For some large p
value, in these experiments, the p-dependent force exerted by the lasers on the atoms changes sign. The
force is thus confining (friction) for small p and expelling for large p. Thus the atoms reaching the expulsion
region are lost from the cooling process, with a probability of one. A specific model with absorbing walls (see
Section 10.3.2) might be introduced to treat this case.

the generalized CLT. For definiteness, consider here eq. (8.7) giving the trapped proportion for the one-dimensional unconfined model in which $\mu = \hat{\mu} = 1/2 < 1$. The generalized CLT tells us that, for large N, the probability densities of the sums $T_N = \sum_{i=1}^{N} \tau_i$ and $\hat{T}_N = \sum_{i=1}^{N} \hat{\tau}_i$ depend *only* on the power-law *tails* of $P(\tau)$ and $\hat{P}(\hat{\tau})$. Thus, just as in the usual CLT where only the first two moments of the density $P(x)$ influence the probability density of $X_N = \sum_{i=1}^{N} x_i$ at large N, here only the tails of $P(\tau)$ and $\hat{P}(\hat{\tau})$ matter, each tail being determined by two parameters, the prefactor and the exponent of the corresponding power-law tail. The reason for the exactness of eq. (8.7) is that our simplifying assumptions are carefully chosen to retain the exact values for the parameters of the tails. Actually, the (generalized) CLT has been used as a guide in Chapter 3 to identify the only relevant parameters at large times. Thus, we have obtained exact asymptotic results starting from simple models of the initially complex laser cooling problems.

All these results support our statistical analysis which gives accurate results in spite of the seemingly oversimplifying assumptions on which it is based. We therefore conjecture that the statistical analysis also gives correct predictions in situations which have not been tested. Note that these situations are much more frequent than those which have already been tested, and this stresses the fact that the statistical analysis constitutes a very powerful tool that allows one to explore situations where no other theoretical tools are available. Since, in addition, it usually yields analytical results, one can easily vary the parameters, and therefore use it in order to optimize subrecoil cooling, as already shown in the case of Raman cooling [RBB95, RSC01]. Taking a different perspective [SSY97] (see also [MZL96]), the results of the Lévy statistics analysis could also be used as an indication of the existence of scaling laws that might remain to be discovered in the microscopic quantum optics evolution laws.

9

Example of application: optimization of the peak of cooled atoms

9.1 Introduction

The statistical approach presented in this book provides not only a deeper physical understanding of subrecoil cooling, but also analytical expressions for the various characteristics of the momentum distribution of the cooled atoms. A great confidence in the validity of these predictions has been obtained in the previous chapter, by comparing them with experimental and numerical results. Therefore, we are now entitled to apply the approach developed in this work to specific problems, such as the optimization of one particular feature of the cooling process, namely the height of the peak of cooled atoms. This is the subject of this chapter.

Finding empirically the optimum conditions for a subrecoil cooling experiment is a difficult task. There are *a priori* many parameters to be explored and each experiment with a given set of parameters is in itself lengthy. The same can also be said of numerical simulations. One needs guidelines such as those provided by the present statistical approach to reduce the size of the parameter space to be explored.

There is a variety of optimization problems that can be considered. Following usual motivations of laser cooling, like the increase of atomic beam brightness or the search for quantum degeneracy, we will concentrate here on optimizing the height $h(\theta)$ of the peak of the momentum distribution of the cooled atoms, which corresponds also to the gain in phase space provided by the cooling (see Section 6.2.3).

We will take as a constraint for the optimization the finiteness of the time θ available for the cooling. In practice, this is the limiting factor in most experiments: after some time, the residual motion of the atoms makes them leave the zone of interaction with the lasers. From the theoretical point of view, the finiteness of the available time θ is fundamental since, as shown above, when $\langle \tau \rangle$ or $\langle \hat{\tau} \rangle$ are infinite, it is the experimental parameter θ that sets the time scale of the phenomenon.

124

Other constraints, like a finite number of available spontaneous photons, could be considered as well[1]. We also focus here on the following aspects.

- We restrict ourselves to situations with friction forces outside the trapping region, which correspond to the confined recycling model (see Section 3.2). Indeed, it is obvious that these confining forces, which are available for both VSCPT and Raman cooling, reduce the return times to the trap and are thus advantageous for the cooling.

- Only the case of $\alpha = 2$ is considered for the jump rate $R(p) \propto p^{\alpha}$ (see eq. (3.5)). This is the case for VSCPT and, in Raman cooling, $\alpha = 2$ has been shown to lead to more efficient cooling than $\alpha = 4$ in one dimension. Optimizing the value of α, which is possible in practice for Raman cooling only, is a problem in itself which is not considered here.

- The jump rate $R(p)$ is taken as constant in time, as everywhere else in this book. Removing this restriction can enlarge the cooling possibilities and lead to yet better optimization schemes; jump rates varying in time would, however, require a significant generalization of the present statistical approach.

- We consider only the one-dimensional case, but previous chapters provide the information for carrying out similar two- and three-dimensional optimizations.

In this chapter, we first use the insight provided by the statistical approach to isolate the relevant parameters of the cooling and find that there is actually only one (Section 9.2). We then explain qualitatively why an optimum arises for the height $h(\theta)$ (Section 9.3). The optimum parameter is first derived in Section 9.4 using the analytical expression for the height. A second, more elegant, derivation of the optimum, based on the properties of Lévy sums is presented in Section 9.5. The important features of the optimized cooling are inferred in Section 9.6. Finally, a specific property of the optimized cooling is exhibited and interpreted in terms of a random walk (Section 9.7).

Some results on the cooling optimization of Raman cooling have already been presented in [RBB95, Rei96].

---

[1] This could soon become relevant for VSCPT cooling experiments with metastable helium on the $2^3S_1 \longleftrightarrow 2^3P_1$ transition. Indeed, the 2^3P_1 state has a small probability of relaxing to the 1^1S_0 ground state, at a rate of 2×10^2 s^{-1} (see [DrD69, LJD77, TaH72], and Section II.2.1.3 in [Bar95] for a review), compared to 10^7 s^{-1} for $2^3P_1 \rightarrow 2^3S_1$. Thus, after exchanging on average $10^7/(2 \times 10^2) = 5 \times 10^4$ photons on the right transition, the atoms will fall to the 1^1S_0 state and be lost to the cooling. This gives rise to the problem of cooling optimization under the constraint of a maximum number of available photons. On the other hand, the finite lifetime of the 2^3S_1 state, $\simeq 8000$ s, is large enough to be of no concern up to now.

9.2 Parametrization

To perform the optimization, one needs to identify the relevant parameters. The statistical analysis has shown that all the physical quantities are derived from the probability distributions of trapping times, $P(\tau)$, and of first return times, $\hat{P}(\hat{\tau})$. One must therefore examine, for each cooling mechanism, on which parameters these distributions depend. For this purpose, we use the expressions of the parameters of $P(\tau)$ and $\hat{P}(\hat{\tau})$ obtained in Appendix A.

Consider first (one-dimensional) VSCPT. The tail of the distribution $P(\tau)$ of trapping times is the only part of $P(\tau)$ that is relevant at large interaction times. It is characterized by the exponent $\mu = 1/2$, which cannot be changed, and by the time parameter τ_b (see eq. (A.41))

$$\tau_b = \frac{\pi}{256} \left(\frac{\Omega_1}{\Omega_R}\right)^2 \left(\frac{\hbar k}{p_{\text{trap}}}\right)^2 \Gamma^{-1}.$$

The distribution of return times $\hat{P}(\hat{\tau})$ is characterized by its average value $\langle\hat{\tau}\rangle$ (cf. confined model), given by eq. (A.48)

$$\langle\hat{\tau}\rangle = 2 \frac{p_{\text{max}}}{p_{\text{trap}}} \left(\frac{\Gamma}{\Omega_1}\right)^2 \Gamma^{-1}.$$

The two above expressions depend on the following three types of parameters.

- The trap size p_{trap} is an arbitrary intermediate parameter that disappears at the end of the calculations and is thus not involved in the optimization.
- The parameters $\hbar k$, Ω_R and Γ are determined by the atomic element and by the atomic levels used for the cooling. For a given cooling problem, these parameters take well defined values that cannot be tuned. The parameter p_{max} that only appears in $\langle\hat{\tau}\rangle$ should obviously be as small as possible. We assume that the laser detuning is adjusted so as to minimize p_{max}, to reach the value given by eq. (A.33)[2].
- The light intensity, proportional to Ω_1^2. This is the *only parameter that can actually be varied in a given experimental setup*. We therefore define the dimensionless optimization parameter λ by

$$\lambda = \left(\frac{\Omega_1}{\Gamma}\right)^2. \tag{9.1}$$

The above equations giving τ_b and $\langle\hat{\tau}\rangle$ can be rewritten as a function of λ:

$$\tau_b = \left(\frac{\hbar k}{p_{\text{trap}}}\right)^2 \frac{\lambda}{Q} \Gamma^{-1}; \tag{9.2a}$$

---

[2] One can vary the laser detuning $\bar{\delta}$ to achieve this; in this case, the expressions presented here must be adapted slightly to include a non-zero $\bar{\delta}$.

$$\langle \hat{\tau} \rangle = \frac{\hbar k}{p_{\text{trap}}} \frac{\hat{Q}}{\lambda} \Gamma^{-1}, \tag{9.2b}$$

where Q and $\hat{Q}$, called the 'atomic parameters', are determined by the specific cooling configuration. These parameters are given in table 9.1.

We now turn to (one-dimensional) Raman cooling with the sequence of time square pulses defined in Section A.2.1. This cooling problem, very similar to VSCPT, has a trapping time exponent $\mu = 1/2$ and a trapping time parameter τ_b given by eq. (A.73):

$$\tau_b = \frac{3\pi^3}{32\Omega_R^2 \tau_{p,1}} \left(\frac{\hbar k}{p_{\text{trap}}} \right)^2.$$

The return time distribution is characterized by (see eq. (A.75)):

$$\langle \hat{\tau} \rangle = \frac{12\tau_{p,1}}{5} \frac{p_{\text{max}}}{p_{\text{trap}}}.$$

These expressions involve the same type of parameters as for VSCPT. There is again *only one parameter that can be varied* and that controls both $P(\tau)$ and $\hat{P}(\hat{\tau})$: $\tau_{p,1}$, the duration of the longest pulses of the sequence. Defining the optimization parameter λ for Raman cooling by

$$\lambda = \frac{1}{\Gamma \tau_{p,1}}, \tag{9.3}$$

τ_b and $\langle \hat{\tau} \rangle$ can be written in the same form as for VSCPT, that is eq. (9.2a) and eq. (9.2b), the corresponding 'atomic parameters' Q and $\hat{Q}$ being given in table 9.1.

The parametrization of eq. (9.2) implies that the distributions $P(\tau)$ and $\hat{P}(\hat{\tau})$ no longer appear as mathematically independent, as was the case up to now. The

Table 9.1. *Optimization parameter λ and atomic parameters Q and $\hat{Q}$. Note the similarity between VSCPT and Raman cooling for the expressions of Q and $\hat{Q}$.*

	1D VSCPT	1D Raman cooling
λ	$(\Omega_1/\Gamma)^2$	$\frac{1}{\Gamma\tau_{p,1}}$
Q	$\frac{256}{\pi} \left(\frac{\Omega_R}{\Gamma} \right)^2$	$\frac{32}{3\pi^3} \left(\frac{\Omega_R}{\Gamma} \right)^2$
$\hat{Q}$	$2 \frac{p_{\text{max}}}{\hbar k}$	$\frac{12}{5} \frac{p_{\text{max}}}{\hbar k}$

physical situations from which they arise are such that they both depend on the same parameter λ (and, as a consequence, can only be varied in a correlated way).

A second important outcome of the parametrization is that the typical trapping time τ_b is an *increasing* function of λ whereas the typical first return time $\langle \hat{\tau} \rangle$ is a *decreasing* function of λ (see also Section A.1.2.2, p. 156 and Section A.2.2.1, p. 169). This will be at the origin of the existence of an optimum (see Section 9.3).

Strikingly, although VSCPT and Raman cooling arise from different physical effects, the same dependences on λ are obtained for typical trapping and return times. This property allows us, when specific calculations are required, to treat explicitly only VSCPT. The Raman cooling case can be inferred following exactly the same method.

9.3 Why is there an optimum parameter?

The height $h(\theta)$ of the peak of cooled atoms is proportional to the number of returns to the trap, as explained in Section 6.2.3. To maximize $h(\theta)$, one must therefore obtain the largest possible number of returns. An optimum will exist if the number of returns has an upper bound when the optimization parameter λ varies. We now explain intuitively why such an upper bound exists, before calculating it by two different methods in the next two sections.

Consider first a very small value of λ. Then, according to eq. (9.2b), the average return time to the trap, $\langle \hat{\tau} \rangle$, will be long and the recycling of the atoms out of the trap will be poor. On the other hand, when the atoms return to the trap (but do not land exactly in $p = 0$), they will relatively rapidly come out of it because the typical trapping time, determined by τ_b, is short (see eq. (9.2a)). The filtering is thus efficient. But, in itself, this does not ensure an efficient cooling because each (short) trapping event is necessarily followed by a (long) return event. In the case of a too small λ, the limiting factor for the number of returns is the too long typical return time. The combination of poor recycling and efficient filtering leads to the accumulation of most atoms outside the trap (see Fig. 9.1) which is detrimental to the height $h(\theta)$ of the cooled peak[3].

In the case of a very large value of λ, the opposite behaviour appears. The average return time $\langle \hat{\tau} \rangle$ is now small, providing efficient recycling. But, when the atoms return to the trap, their typical trapping time is long, even if they are not very close to $p = 0$. Paradoxically, this is not good for the cooling either because now the filtering is inefficient and atoms tend to spend too much time in the trap. The long typical trapping times are now the limiting factor for the number of returns

---

[3] This is true up to a certain time scale depending on λ. For asymptotically long times, the filling of the trap is given by the results of Chapter 5. This stresses the fact that the optimization discussion involves a finite (yet large) number N of trapping events, a point which will be made clearer in Section 9.7.

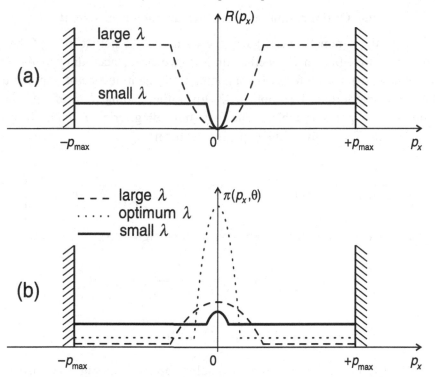

Fig. 9.1. *Intuitive explanation for the existence of a cooling optimum.* (a) Jump rate $R(p_x)$ for a small λ and for a large λ (see Fig. A.5, p. 157). (b) Corresponding schematic momentum distributions $\pi(p_x, \theta)$. When λ is too small, the atoms spend too much time in the recycling region. When λ is too large, they spend too much time in the trapping region. The optimum momentum distribution, yielding the highest cooled peak, is obtained for an intermediate value of λ, calculated in Sections 9.5 and 9.6.

(see also Section 7.1.4). The combination of efficient recycling and poor filtering leads to the accumulation of most atoms in the trap, but not in the close vicinity of $p = 0$ (see Fig. 9.1). This is again detrimental for the height $h(\theta)$ of the cooled peak.

Between these two extreme cases of a too small or of a too large λ, there must exist an optimum regime for which the interaction time θ is adequately shared between trapping and recycling. The competition between trapping and recycling (see Section 3.1) must not be completely dominated by trapping, as one might naively have believed. What will come out of the optimization calculation is that the atoms must spend approximately the same time in the trap and out of the trap (see Section 9.7).

9.4 Optimization using the expression of the height

To compute the parameter $\lambda_{opt}(\theta)$ which maximizes the height $h(\theta)$ of the peak of cooled atoms after an interaction time θ, we need, according to the above discussion, an expression for $h(\theta)$ containing both the influence of trapping terms and of recycling terms. Equation (6.32) gives the leading term of $h(\theta)$, which contains only trapping variables. The first subleading term can be derived from eqs. (6.23) and (5.20), giving the Laplace transform[4]

$$\mathcal{L}h(s) = \frac{1}{V_D(p_{trap})} \left[\frac{1}{\Gamma(1-\mu)\tau_b^\mu} \frac{1}{s^{1+\mu}} - \frac{\langle\hat{\tau}\rangle}{\left(\Gamma(1-\mu)\tau_b^\mu\right)^2} \frac{1}{s^{2\mu}} + \cdots \right], \quad (9.4)$$

from which we infer the large-θ behaviour of the height $h(\theta)$[5]

$$\begin{aligned} h(\theta) = \frac{1}{V_D(p_{trap})} \Big[&\frac{1}{\Gamma(1-\mu)\Gamma(1+\mu)\tau_b^\mu} \theta^\mu \\ &- \frac{\langle\hat{\tau}\rangle}{\left(\Gamma(1-\mu)\tau_b^\mu\right)^2 \Gamma(2\mu)} \theta^{2\mu-1} + \cdots \Big]. \end{aligned} \quad (9.5)$$

The leading term ($\propto \theta^\mu$) is determined solely by the trapping time distribution $P(\tau)$, which confirms that trapping dominates the phenomenon. On the contrary, the first subleading term ($\propto \theta^{2\mu-1}$) of eq. (9.5) depends on return times through $\langle\hat{\tau}\rangle$. This negative term reduces the height of $h(\theta)$, and this reduction increases with $\langle\hat{\tau}\rangle$. This is consistent with the intuitive fact that, for a given τ_b, the longer the time spent out of the trap, the smaller the height $h(\theta)$ of the cooled peak. *In short, eq. (9.5) represents very clearly the competition between trapping (positive leading term) and recycling (negative subleading term) (cf. Section 3.1) described qualitatively in the previous section.*

The time scale beyond which this equation is valid, i.e. beyond which the subleading term becomes smaller than the leading term, is easily calculated to be the same scale θ_0 as that defining the validity of the expansion (5.23) for $f_{trap}(\theta)$ (see eq. (5.24)). This is consistent with the fact that eq. (9.5) and eq. (5.23) rely on the same assumption that the typical trapping times dominate the typical return times, a condition which is achieved for $\theta > \theta_0$ (see Section 5.2.4).

---

[4] In eq. (5.20) we have neglected $A_0\tau_b$ in comparison to $\langle\hat{\tau}\rangle$, since τ_b is of the order of τ_{trap} which is much smaller than the average return time in the trap.

[5] One can check that the first term of eq. (9.5) is identical to eq. (6.32), noting that $V_D(p_{trap}) = C_D p_{trap}^D$ (see eq. (3.24)), $\Gamma(1-\mu)\Gamma(1+\mu) = \mu\Gamma(1-\mu)\Gamma(\mu) = \pi\mu/\sin(\pi\mu)$ (see eq. (4.47)), $\tau_b^\mu = A_\mu \tau_{trap}^\mu$ (see eq. (3.33)) and $\theta^\mu/(p_{trap}^D\tau_{trap}^\mu) = 1/p_\theta^D$ (cf. eqs. (6.2) and (3.31)).

We can now introduce the expressions of τ_b and $\langle \tau \rangle$ parametrized by λ (eq. (9.2a) and eq. (9.2b)), in the case $\mu = 1/2$:

$$h(\theta) = \frac{Q^{1/2}}{2\pi\hbar k} \left[\frac{2\,(\theta\Gamma)^{1/2}}{\lambda^{1/2}} - \frac{Q^{1/2}\hat{Q}}{\lambda^2} \right] \tag{9.6}$$

(we recall that $\Gamma(1/2) = \sqrt{\pi}$, $\Gamma(1) = 1$). Note that the reduction of $h(\theta)$ by the subleading term does not depend on θ, a feature specific of the case $\mu = 1/2$. Note also that, as expected, the intermediate parameter p_{trap} no longer appears in this expression.

The parameter $\lambda_{\text{opt}}(\theta)$ that maximizes $h(\theta)$ at the end of a given interaction time θ satisfies $\partial h(\theta)/\partial\lambda = 0$, which gives

$$\lambda_{\text{opt}}(\theta) = \frac{2^{2/3} Q^{1/3} \hat{Q}^{2/3}}{(\theta\Gamma)^{1/3}}. \tag{9.7}$$

One can check in eq. (9.6) that $\lambda_{\text{opt}}(\theta)$ is in the regime where the subleading term is smaller than the leading term. However, the two terms are found to differ only by a factor of four, so that the expansion of $h(\theta)$ is valid but maybe not very accurate. We will give an alternative, more direct, derivation of the above result in the next section.

The optimum of the height $h(\theta)$ has been studied numerically [Rei96] in the case of Raman cooling using an appropriate Monte Carlo simulation. The numerically found optimum agrees satisfactorily, to better than 50%, with the prediction of eq. (9.7).

9.5 Optimization using Lévy sums

We have noted that the expression (9.5) of $h(\theta)$, from which we performed the optimization, is based on the assumption that the additional subleading terms are negligible (typical trapping time $\gg$ typical recycling time). This gives an optimum situation which satisfies this assumption only marginally. This assumption was legitimate in Chapters 5 and 6 when studying the asymptotic time regime, for a given parameter λ. However, it is possible that, for a given time θ, the assumption is not satisfied when taking $\lambda = \lambda_{\text{opt}}(\theta)$. If this were the case, the calculation presented in Section 9.4 would be questionable.

We now present a more elegant, direct method that fully confirms the above result. It is based directly on the properties of Lévy sums presented in Chapter 4. It only relies on the fact that the number N of return events is large and, actually, does not even require the results of Chapters 5 and 6. It does not make any assumptions about the relative values of trapping times versus recycling times.

We begin with eq. (3.3) writing the available time θ as the sum of the total trapping time T_N and the total recycling time $\hat{T}_N$:

$$\theta \simeq T_N + \hat{T}_N,$$

where N is the number of return events. We have established in Section 6.2.3 that the peak height $h(\theta)$ is proportional to the number N of returns. Therefore the optimization problem now amounts to finding the parameter λ, maximizing N with the above equation as a constraint.

According to the generalized Central Limit Theorem (CLT) (cf. eq. (4.9)), for $\mu = 1/2 < 1$, the Lévy sum T_N behaves as

$$T_N \simeq N^2 \tau_{\mathrm{b}}, \tag{9.8}$$

for $N \gg 1$. According to the usual CLT (cf. eq. (4.4)), as $\langle \hat{\tau} \rangle$ is finite, the Lévy sum $\hat{T}_N$ behaves as

$$\hat{T}_N \simeq N \langle \hat{\tau} \rangle, \tag{9.9}$$

for $N \gg 1$. Injecting these two expressions into eq. (3.3), and using eqs. (9.2a) and (9.2b) for τ_{b} and $\langle \hat{\tau} \rangle$, we obtain

$$\theta \simeq N^2 \left(\frac{\hbar k}{p_{\mathrm{trap}}} \right)^2 \frac{\lambda}{Q} \, \Gamma^{-1} + N \, \frac{\hbar k}{p_{\mathrm{trap}}} \, \frac{\hat{Q}}{\lambda} \, \Gamma^{-1}. \tag{9.10}$$

Equation (9.10) has a single positive solution:

$$N = \frac{Q\hat{Q}}{2} \, \frac{p_{\mathrm{trap}}}{\hbar k} \, \frac{-1 + \sqrt{1 + 4 \, \frac{\Gamma\theta}{Q\hat{Q}^2} \, \lambda^3}}{\lambda^2}. \tag{9.11}$$

This function first grows and then decreases as a function of λ. The position of the maximum is given by:

$$\lambda_{\mathrm{opt}}(\theta) = 2^{1/3} \left(\frac{Q\hat{Q}^2}{\Gamma\theta} \right)^{1/3}. \tag{9.12}$$

This corresponds to the approximate result of eq. (9.7), within a factor of $2^{1/3} \simeq 1.26$. Thus the optimum derived in Section 9.4 and its subsequent properties (see Sections 9.6 and 9.7) are valid. Moreover, this short new derivation demonstrates how the generalized CLT can lead to useful results in a few steps.

Note finally that, for the optimum value of λ, the total trapping time T_N and the total return time $\hat{T}_N$ are actually both roughly equal to $\theta/2$. This important result clearly shows that the optimum of the peak height is obtained when the time θ is shared equally between trapping and recycling. This result was not obvious *a priori*.

9.6 Features of the optimized cooling

The optimum parameter $\lambda_{\text{opt}}(\theta)$ of eq. (9.7) depends not only on the atomic parameters Q and $\hat{Q}$ but also, interestingly, *on the chosen interaction time θ*. Therefore, the cooling features obtained in Chapter 6 as functions of θ for a given λ (independent of θ) will acquire an additional time dependence if we choose $\lambda = \lambda_{\text{opt}}(\theta)$. For practical purposes, it is these optimized quantities that will be relevant to estimate the potentialities of a given experimental situation. Here we calculate the expressions for the optimized quantities and discuss their new time dependences, as compared with the unoptimized quantities found in Chapter 6. For definiteness, we establish and discuss these expressions for VSCPT. The results are the same for Raman cooling, provided one makes the right substitutions (see table 9.1).

By substituting into eq. (9.7) the expressions for Q and $\hat{Q}$ given in table 9.1, we get

$$\lambda_{\text{opt}}(\theta) = \frac{16}{\pi^{1/3}} \left(\frac{p_{\text{max}}}{\hbar k} \frac{\Omega_{\text{R}}}{\Gamma} \right)^{2/3} \frac{1}{(\theta\Gamma)^{1/3}}. \tag{9.13}$$

Applying eq. (9.1), we have the Rabi frequency $\Omega_{1,\,\text{opt}}$ optimized for an experiment of duration θ:

$$\Omega_{1,\,\text{opt}} = \frac{4}{\pi^{1/6}} \left(\frac{p_{\text{max}}}{\hbar k} \frac{\Omega_{\text{R}}}{\Gamma} \right)^{1/3} \frac{1}{(\theta\Gamma)^{1/6}} \, \Gamma. \tag{9.14}$$

The corresponding characteristic momentum p_θ is then calculated, at the time θ for which the cooling is optimized, by substituting $p_0\tau_0^{1/2}$ (see eq. (A.25) with Ω_1 given by eq. (9.14)) into eq. (6.2):

$$\begin{aligned}
\left. \frac{p_\theta}{\hbar k} \right|_{\text{opt}} &= \frac{1}{\pi^{1/6}} \left(\frac{p_{\text{max}}}{\hbar k} \right)^{1/3} \left(\frac{\Omega_{\text{R}}}{\Gamma} \right)^{-2/3} \frac{1}{(\theta\Gamma)^{1/6} (\theta\Gamma)^{1/2}} \\
&= \frac{1}{\pi^{1/6}} \left(\frac{p_{\text{max}}}{\hbar k} \right)^{1/3} \frac{1}{(\theta\Omega_{\text{R}})^{2/3}}.
\end{aligned} \tag{9.15}$$

Thus the half-width $p_{\theta,\text{opt}}$ decreases significantly faster with θ than the unoptimized half-width ($\propto \theta^{-1/2}$), due to the additional factor $\theta^{-1/6}$ coming from the optimization. In other words, thanks to the optimization, *the cooling possibilities improve more rapidly with the interaction time than could be anticipated from the initial analysis assuming a given Ω_1*. This is an important outcome of the optimization.

Moreover, the optimized half-width depends on the wall position p_{max}, while the unoptimized p_θ (see eq. (6.2)) did not depend on it. However, this dependence is weak ($p_{\theta,\text{opt}} \propto p_{\text{max}}^{1/3}$). Thus, for practical purposes, the precise value of p_{max} is not at all critical. This non-trivial result is qualitatively confirmed by numerical calculations [MDT94] which showed that friction in one-dimensional VSCPT (finite

p_{max}) does not lead, for relatively modest interaction times ($\theta \simeq 2 \times 10^4 \Gamma^{-1}$), to a dramatic improvement of the cooling compared with one-dimensional VSCPT without friction (infinite p_{max}).

It is instructive to compare these optimum conditions with those of the most advanced VSCPT experiments [SLC99]. Although the present sophisticated analysis was not available at that time, those experiments had to be at least grossly optimized in order to obtain a sufficient signal of cooled atoms. They were performed in the time range $\theta = 2 \times 10^3 - 1.5 \times 10^4 \Gamma^{-1}$. Let us take $\Omega_R / \Gamma = 1/37.44$, the metastable helium value, and $p_{max} = 10\hbar k$, a reasonable order of magnitude for the maximum momentum reached by the atoms for these experiments without friction. The range given by eq. (9.14) for the optimum Rabi frequency[6] is then $\Omega_{1, opt} = 0.43 - 0.6\Gamma$. This is reasonably close to the chosen fixed value for these experiments, $\Omega_1 \simeq 0.72(2)\Gamma$.

The optimized height $h(\theta)|_{opt}$ of the cooled peak, at the time θ for which it is optimized, can be derived by substituting p_θ of eq. (9.15) into eq. (6.32):

$$\left. \frac{h(\theta)}{(\hbar k)^{-1}} \right|_{opt} = \frac{2\hbar k}{\pi^{3/2} p_\theta} = \pi^{-5/6} \left(\frac{p_{max}}{\hbar k} \right)^{-1/3} \left(\frac{\Omega_R}{\Gamma} \right)^{2/3} (\theta \Gamma)^{1/6} (\theta \Gamma)^{1/2}$$

$$= \pi^{-5/6} \left(\frac{p_{max}}{\hbar k} \right)^{-1/3} \left(\frac{\Omega_R}{\Gamma} \right)^{2/3} (\theta \Gamma)^{2/3}. \qquad (9.16)$$

The increase of $h(\theta)$ with θ is significantly faster than it was for the unoptimized solution ($\propto \theta^{1/2}$) due to the additional factor $\theta^{1/6}$ coming from the optimization and the dependence on p_{max} is weak.

Finally, it is interesting to analyse how the trap fills in the optimized conditions. We first introduce the expressions (9.2a) and (9.2b) into eq. (5.23) for $f_{trap}(\theta)$, then substitute the optimum value $\lambda_{opt}(\theta)$ of eq. (9.7). We obtain the trapped proportion at the time θ for which the peak height is optimized:

$$\left. f_{trap}(\theta) \right|_{opt} = 1 - \frac{1}{\pi} \frac{Q^{1/2} \hat{Q}}{\left(\lambda_{opt}(\theta) \right)^{3/2} (\theta \Gamma)^{1/2}} = 1 - \frac{1}{2\pi}. \qquad (9.17)$$

This expression for the time dependence of the trapped population under optimum peak height is remarkable in that it does not depend at all on any atomic (Γ, Ω_R, $\hbar k$) or experimental (p_{max}, Ω_1) parameter. It is a universal result, provided τ_b and $\langle \hat{\tau} \rangle$ can be parametrized as in eqs. (9.2a) and (9.2b).

Moreover, eq. (9.17) reveals that optimizing the peak height at θ does not lead to the maximum possible value of one for $f_{trap}(\theta)$, which could be obtained by taking $\Omega_1 > \Omega_{1, opt}$, but rather $f_{trap}(\theta)|_{opt} = 1 - \frac{1}{2\pi} = 0.84 \ldots$. This is because the peak height optimization results from a compromise (see Section 9.3) in which

---

[6] These values are slightly underestimated because the condition $\Omega_1 \ll \Gamma/2$ used to derive the expression eq. (A.23) for τ_0 is obviously not well satisfied here. See footnote 2 in Appendix A, p. 151.

the typical trapping time must not be made too large in order not to slow down the diffusion process that eventually leads the atoms to $p = 0$. We however note that $f_{\mathrm{trap}}(\theta)$ is not too far from one, which will be qualitatively explained in the following section.

9.7 Random walk interpretation of the optimized solution

The optimum cooling condition (9.7) implies a very specific regime for the random walk performed by the atoms. To understand this feature, let us calculate, for the optimum condition $\lambda = \lambda_{\mathrm{opt}}(\theta)$, the average return time $\tau_{\mathrm{ret}}(\theta)|_{\mathrm{opt}}$ to the peak of half-width $p_{\theta,\,\mathrm{opt}}$, more specifically to a dark state of momentum less than $p_{\theta,\,\mathrm{opt}}$. This return time is given by the same expression (9.2b) as $\langle \hat{\tau} \rangle$ except that p_{trap} must be replaced by $p_{\theta,\,\mathrm{opt}}$, and that there is an additional factor of two due to the fact that the atoms returning to $p < p_{\theta,\,\mathrm{opt}}$ have a probability $1/2$ of falling into a dark state (see Section A.1.3):

$$\tau_{\mathrm{ret}}(\theta)|_{\mathrm{opt}} = 2 \, \frac{\hbar k}{p_{\theta,\,\mathrm{opt}}} \, \frac{\hat{Q}}{\lambda_{\mathrm{opt}}(\theta)} \, \Gamma^{-1}. \tag{9.18}$$

Using table 9.1, eq. (9.13) and eq. (9.15), the following expression is obtained:

$$\tau_{\mathrm{ret}}(\theta)|_{\mathrm{opt}} = \frac{\pi^{1/2}}{4} \, \theta = 0.44 \ldots \theta. \tag{9.19}$$

Thus, in the optimum configuration, the return time $\tau_{\mathrm{ret}}(\theta)|_{\mathrm{opt}}$ to the peak is of the order of the total time θ. This means that, when optimizing the peak height, we set at the same time for the atoms a momentum 'target', the half-width $p_{\theta,\,\mathrm{opt}}$. Equation (9.19) shows indeed that the available time θ is just about the required time $\tau_{\mathrm{ret}}(\theta)$ for the atoms to reach the peak[7].

As a consequence of eq. (9.19), we expect the peak to contain a significant fraction of the atoms, which is precisely confirmed by eq. (9.17). Thus, *by optimizing the peak height, nearly all the atoms are brought into the cooled peak.* In other words, optimizing the *height* of the peak of the cooled atoms also corresponds to nearly optimizing the *cooled fraction*. This is a valuable feature for experiments: these optimum conditions correspond to a large signal of cooled atoms. This property had first been proposed in [RBB95] as a heuristic method to derive the optimum *height*: the optimum was calculated by assuming that the jump rate $R(p)$ should ensure $\tau_{\mathrm{ret}}(\theta)|_{\mathrm{opt}} \simeq \theta$.

Yet, this approximate compatibility between the *optimum height* and the *optimum cooled fraction* is far from being *a priori* obvious. The height optimum could

---

[7] Note that, on the contrary, the required time to reach $p < p_{\mathrm{trap}}$ is much smaller than θ since p_{trap} is usually much larger than $p_{\theta,\,\mathrm{opt}}$. This ensures that, in spite of eq. (9.19), the number N of returns to the trap is large, which is essential when applying our statistical analysis.

conceivably correspond to a very narrow cooled peak containing very few atoms, yet optimally high, which would be much less interesting for experiments. The fact that the optimization of both the peak height and the cooled fraction happen simultaneously is certainly one of the reasons for the success of the one-dimensional VSCPT and Raman cooling experiments.

10

Conclusion

10.1 What has been done in this book

In this book, we have introduced and developed a model for subrecoil cooling, which is the scheme leading to the lowest temperatures achieved today by laser cooling. This model is inspired by the quantum jump descriptions of the cooling process and represents the evolution of the atom in terms of an inhomogeneous random walk in momentum space with a momentum-dependent jump rate $R(p)$ which vanishes for $p = 0$.

We have shown that such an inhomogeneous random walk gives rise to broad distributions. More precisely, the distribution $P(\tau)$ of the trapping times τ of the atom in a small zone around $p = 0$ has power-law tails for $\tau \to \infty$ which can decrease so slowly with τ that the mean value and/or the variance of τ diverge. Similar broad distributions can also exist for the distribution $\hat{P}(\hat{\tau})$ of the first return times $\hat{\tau}$ of the atom to the trapping zone.

Lévy statistics is the relevant framework for the analysis of problems involving sums of random variables with broad power-law distributions. We have thus presented a brief review of Lévy statistics and introduced a 'sprinkling distribution' from which we have been able to derive analytical expressions for important physical quantities characterizing the cooling process, such as the proportion of cooled atoms or the momentum distribution.

Finally, we have compared these analytical predictions with the results of experimental investigations or numerical simulations. We have shown that, whenever quantitative experimental or numerical results are available, excellent agreement is obtained with our analytical predictions. We have also used our statistical approach for studying the cooling optimization.

10.2 Significance and importance of the results

10.2.1 From the point of view of Lévy statistics

There is an increasing number of problems in various fields (biology, physics, finance, etc.) where broad power-law distributions are identified and where, consequently, Lévy statistics plays an important role.

The advantage of the physical problem considered in this work (subrecoil cooling) is that all the parameters appearing in the broad distributions $P(\tau)$ and $\hat{P}(\hat{\tau})$ are related to well defined physical quantities that are well known (atomic mass, atomic frequency, radiative lifetime) or which can be experimentally measured (Rabi frequency). Furthermore, for certain laser configurations, a precise comparison with exact results, derived from a microscopic quantum optics approach, can be made. Subrecoil cooling is therefore an example of a physical problem allowing a precise and quantitative test of theories constructed from Lévy statistics.

In fact, most of the mathematical equations used in this book for calculating physical quantities, such as the cooling efficiency or the momentum distribution, do not use the Lévy distributions themselves, but rather the sprinkling distribution $S(t)$. Our approach therefore emphasizes the importance of such a distribution $S(t)$. We have also shown how the unexpected properties of $S(t)$ when either $\langle \tau \rangle$ or $\langle \hat{\tau} \rangle$ is infinite are crucially important for laser cooling. As mentioned in Section 4.4.2, the sprinkling distribution is known in the context of stochastic processes as the density of a renewal process, or the 'renewal density'. A theory of renewal processes *with infinite mean values* was developed in the 1960's [Dyn61, Lam58] and was considered a mathematical achievement (see Section XIV.3, p. 472 in [Fel71]). However, this theory was considered academic, without any applications to natural sciences, because of the apparent irrelevance of problems involving infinite mean values. The present work on laser cooling, in which we find infinite mean values and where we have been led to recreate independently a basic backbone of renewal theory with infinite mean values, has reactivated mathematical interest in these renewal theories. Some of our results have been confirmed and some have been made more precise within this adequate mathematical setting [BaB00].

It has also been noted [BBJ00] that the approach presented here poses the problem of the role of a singularity in Markov processes, the singularity in subrecoil cooling being created by the vanishing of the excitation rate $R(p)$ at $p = 0$. This rich question of Markov processes with singularities had apparently not been raised before within the mathematical community. Ongoing work in this field [BBJ00], stimulated by the present approach, could in turn have implications for laser cooling and other physics problems of diffusion in the presence of trapping states.

Thus, the physical problems investigated in this book contain several features which are related to recent mathematical developments of the theory of stochastic processes and which stimulate further mathematical work.

10.2.2 From the point of view of laser cooling

The Lévy statistics approach presented in this book yields analytical expressions for the momentum distribution of atoms cooled below the recoil limit. These expressions are, furthermore, in excellent agreement with both experimental and numerical observations. Compared with the usual methods of quantum optics, which look very difficult to apply to subrecoil cooling, statistical methods using Lévy statistics thus appear to be remarkably efficient.

The present approach leads in addition to new physical insights into the cooling process. It identifies the important physical parameters, such as the exponent α of the p-dependence of the jump rate $R(p)$ near $p = 0$, or the ratio $\mu = D/\alpha$ between the dimensionality D and α. It emphasizes also basic features of subrecoil cooling, such as its non-ergodicity.

Finally, the fact that we have obtained analytical expressions for the various quantities characterizing the cooled atoms allows for an optimization of the cooling process. Two examples of such an optimization have been mentioned.

(i) The first is related to subrecoil Raman cooling where the exponent α can be varied by choosing appropriate shapes for the laser pulses, square pulses ($\alpha = 2$) or Blackman pulses ($\alpha \simeq 4$). The calculated α-dependence of the width of the momentum distribution allows one to predict that square pulses (which are simpler to make than Blackman pulses) are more efficient for one-dimensional Raman cooling, since they lead to a faster decrease of the width of the momentum distribution with the interaction time.

(ii) The second example is the determination of the optimal Rabi frequency Ω_1 in one-dimensional cooling, discussed in Chapter 9. We have shown by two different methods that, for each value of the interaction time θ, there is an optimum Rabi frequency which maximizes the height of the peak of cooled atoms. We have also shown that adjusting the optimal Rabi frequency when θ is varied leads to a faster decrease of the temperature and to a faster increase of the peak height.

The approach followed in this work is therefore useful for guiding the choice of experimental parameters in a given situation. One could also try to use the optimized subrecoil cooling methods in practical applications, such as atomic clocks using ultracold atoms.

10.3 Possible extensions

10.3.1 Improving the optimization

The optimization mentioned in the previous subsection determines the best value of the Rabi frequency Ω_1 for a given value of the interaction time θ, assuming that we keep the same value of Ω_1 between $t = 0$ and $t = \theta$. However, the best values of the cooling parameters are certainly not the same at the beginning of the cooling, when the proportion of cooled atoms is small and when the momentum distribution is broad, and at longer times, when a large fraction of atoms is concentrated in a narrow peak. Thus, one could try to vary continuously Ω_1 over the course of time. That would introduce an extra degree of freedom into the problem and could lead to a significant improvement of the final result for a given interaction time.

10.3.2 More precise model of friction-assisted VSCPT

A friction force can be introduced in VSCPT by a Sisyphus cooling mechanism which operates for a blue detuning of the laser field [SHP93, MDT94, WEO94]. In fact, this mechanism is predominant only for sufficiently weak velocities. At higher velocities, the mean force changes its sign because of a Doppler induced imbalance between opposite radiation pressure forces. This 'anticooling' force, obtained for a blue detuning, is, up to a sign reversal, nothing but the usual Doppler cooling force, obtained for a red detuning. These effects have been experimentally observed [LKS95].

Because of the competition between Sisyphus cooling and Doppler anticooling, the description of the random walk in momentum space should therefore be modified to include these effects. The jump rate $R(p)$ has the same shape for small p. It still vanishes for $p = 0$. But instead of introducing a reflecting wall at $p = p_{max}$, modelling the friction mechanism, one could introduce an *absorbing wall*, which eliminates the atoms if their momentum becomes greater than $p = p_{max}$. A force pushing the atoms towards $p = 0$ if $p < p_{max}$ could be also introduced for modelling more precisely the effect of Sisyphus cooling. This could be done by introducing a dissymmetry between the probability of making a jump putting the atom closer to $p = 0$ or farther away. It is clear that the position of $p = p_{max}$ depends on the Rabi frequency Ω_1, and that the value of Ω_1 would have to be optimized in these new conditions.

10.3.3 Extension to other cooling schemes

In this book we have applied our statistical analysis of an inhomogeneous random walk in momentum space to the two schemes, VSCPT and Raman cooling, that have already led to temperatures well below the recoil limit. However, other schemes that also rely on an inhomogeneous random walk to achieve subrecoil cooling have been proposed [PHB87, WaE89, Mol91] and we think that it might be interesting to examine them in the light of the present analysis. These schemes may apply to atomic species for which VSCPT and Raman cooling are inapplicable.

Among these schemes, the so-called 'broadband Doppler cooling on a narrow transition' [WaE89] is of special interest since it has recently been implemented experimentally [KII99, IIK00, BWS01]. These experiments reached temperatures just above the recoil temperature and led to high phase-space densities.

Broadband Doppler cooling on a narrow transition is an implementation of the standard Doppler cooling scheme using two counterpropagating laser beams detuned below an atomic resonance, with the specific feature that the width of the resonance is narrower than the recoil frequency. A numerical integration of the GOBE shows that subrecoil cooling might be achieved by broadening the laser line and choosing the detuning so that the edge of the spectrum is less than one recoil frequency below resonance.

Although it was first introduced as a variation of the standard Doppler cooling scheme wherein the friction force is the essential ingredient, its specific subrecoil character relies crucially on the inhomogeneity of the random walk in momentum space: the fluorescence rate presents a clear dip around $p = 0$ [Wal95], allowing atoms to accumulate in the vicinity of $p = 0$. (This inhomogeneity is usually negligible in standard Doppler cooling, for which the recoil frequency is smaller than the width of the resonance.)

In contrast to VSCPT and Raman cooling, the fluorescence in broadband Doppler cooling on a narrow transition does not completely vanish at the bottom of the dip. We think, however, that this cooling mechanism can be revisited from the point of view presented in this book. More precisely, the situation for the trapping times reminds one of that studied in Section 7.4 (a non-vanishing jump rate at $p = 0$) and the situation for the recycling times reminds one of the confined model of Section 3.2.3 (a finite average recycling time thanks to Doppler cooling). Applying the kind of analysis developed in this book might allow one to optimize the parameters, thus improving the efficiency of cooling.

10.3.4 Extension to trapped atoms

Introducing a trap for confining the *position* of the atoms brings important advantages. Longer interaction times and higher densities can be obtained because the atomic cloud does not expand ballistically. Achieving a subrecoil cooling in these conditions[1] would be very interesting in the perspective of reaching the threshold

---

[1] Note that the experiments using the broadband Doppler cooling method on a narrow transition, mentioned in Section 10.3.3, were performed in traps.

for quantum degeneracy with purely radiative cooling (without evaporative cooling).

There are, however, a certain number of difficult problems which have to be solved. In the case of VSCPT, for example, there is no longer a perfect dark state. Raman cooling can still be applied to cool trapped atoms below the recoil limit [LAK96] but, up to now, the obtained phase space density has not reached the critical value for quantum degeneracy.

The theoretical description of the atomic random walk in the presence of a trap in position space raises interesting questions. The state of the system between two jumps is no longer described by a single number, the momentum or quasi-momentum p. One must now have two numbers, x and p. From the statistical point of view, the problem has more degrees of freedom and is thus richer than for free atoms. From the point of view of quantum optics, there are also interesting issues. Is it still possible to associate in a rigorous way a classical random walk with the quantum problem? If this is not possible, can we model more precisely the random walk of the state vector of the system in Hilbert space?

Another possibility would be to alternate in time subrecoil cooling phases without trapping, where the results of this work can be applied, and trapping phases preventing the atoms from diffusing too far in space.

10.3.5 Inclusion of many-atom effects

The calculations presented in this book concern the random walk of a single atom. New effects appear when many atoms are involved and the model studied here could be improved to include them. We give two examples of such situations. Note also that atom–atom collisions certainly play an important role (see for instance [GPS95]).

The first example is multiple scattering. The atoms being cooled emit fluorescence photons which are not in the same electromagnetic modes as the photons of the cooling laser beams. These fluorescence photons can therefore be reabsorbed (multiple scattering process) by atoms which are already trapped in quasi-dark states. This introduces a loss mechanism by removing atoms from the trapping states. One could try to model it by introducing a non-zero jump rate near $p = 0$ proportional to the number of jumps made by the atoms which are not in the trap. This might modify qualitatively the cooling behaviour at long times (see for example the discussion presented in Section V.6.3 of [Bar95]).

The second example concerns quantum statistical effects. If the atoms are bosons and if the quasi-dark states have an appreciable population, the probability for the atoms to make a jump which puts them into such states can be enhanced by the Bose factor and this can accelerate the cooling [NWS96]. Determining which of these two processes is dominant as a function of the different parameters is crucial to obtain Bose–Einstein condensation with subrecoil cooling. We believe that such studies can benefit from the statistical approach of the present work.

Appendix A

Correspondence between parameters of the statistical models and atomic and laser parameters

We establish here the correspondence between the statistical models introduced in Chapter 3 and the quantum evolution of atoms undergoing subrecoil laser cooling. This enables us to establish analytical expressions connecting the parameters of the statistical models (τ_0, p_0, p_D, Δp, p_{max}, τ_b and $\hat{\tau}_b$) to atomic and laser parameters relevant to subrecoil laser cooling.

Such a 'dictionary' is useful for the numerical estimation of the results derived in this book (see Chapter 8). It also leads to analytical relations between τ_b and $\hat{\tau}_b$, which are used for cooling optimization (see Chapter 9).

We first treat in detail Velocity Selective Coherent Population Trapping in Section A.1. Analytical expressions are given for the statistical parameters. Special attention is given to the p-dependences of the jump rates both for small p and for large p, because they control the asymptotic behaviours of the trapping and recycling times. It is thus important to include these p-dependences correctly in the simplified jump rates in order to ensure the validity of the statistical model. Raman cooling is then briefly treated in Section A.2.

We only consider here the limit of small laser intensities (and a null detuning for VSCPT) but it is clear that the calculations can easily be generalized if needed.

A.1 Velocity Selective Coherent Population Trapping

We first present the quantum optics treatment of one-dimensional σ_+/σ_- VSCPT (Section A.1.1). This allows us to calculate the exact jump rate, as well as some useful approximations of it, for all values of the momentum p. We also show how the introduction of a fictitious situation with modified atomic and laser parameters leads to a jump rate unchanged in the trapping region, but with a negligible Doppler effect at large values of p.

In Section A.1.2, using the results of Section A.1.1, we express the parameters τ_0, p_0 and p_D as functions of the laser and atomic parameters. We also introduce the elementary step Δp of the random walk in momentum space, and the value $p_{\max}$ allowing us to mimic by confining walls the Sisyphus friction force in the cases when it exists (two-dimensional and three-dimensional VSCPT, one-dimensional lin/45/lin VSCPT).

Section A.1.3 is devoted to calculation of the quantity τ_b, characterizing the *distribution of the trapping times*. The expressions derived for τ_b are valid for all three recycling models (which are identical around $p = 0$).

In Section A.1.4, we express the quantity $\hat{\tau}_b$ characterizing the *distribution of the recycling times*. Here, we find different results for the three different recycling models, which feature different behaviours at large values of p and therefore different recycling properties.

A.1.1 Quantum calculation of the jump rate

We consider the one-dimensional σ_+/σ_- VSCPT scheme for a $J = 1 \rightarrow J = 1$ transition. The results presented here are based on the quantum analysis of [AAK89] with the notation change $\Omega_1 = K_+\sqrt{2} = K_-\sqrt{2}$, and on the delay function quantum analysis of [CBA91] and [Bar95]. Because of optical pumping, the situation can be reduced to a three-level system in a Λ configuration, interacting with two lasers counterpropagating along x, of equal intensities, and circularly polarized σ_+ and σ_-, respectively (Fig. A.1).

Taking into account the quantum description of the atomic motion along the direction x of propagation of the lasers, one finds families $\mathcal{F}_p = \{|g_-\rangle_p, |g_+\rangle_p, |e\rangle_p\}$ of three coupled levels, characterized by a generalized momentum p along x. A given family is stable under the atom–laser interaction. These three states are $|g_\pm\rangle_p = |g_\pm, p \pm \hbar k\rangle$ and $|e\rangle_p = |e, p\rangle$, where $p \pm \hbar k$ and p are the atomic momenta associated with $g_\pm$ and e, respectively. An atom in g_+ (g_-) must have a momentum $p + \hbar k$ $(p - \hbar k)$ to reach the state $|e, p\rangle$ by absorbing a photon σ_- (σ_+) with momentum $-\hbar k$ $(+\hbar k)$. The label p used in the notation $|g_\pm\rangle_p$, $|e\rangle_p$ is actually the momentum of the total system 'atom + laser photons'.

When, in addition, one takes into account the fluorescence process in which an atom in a ground state $|g_\pm\rangle_p$ absorbs a laser photon and re-emits it spontaneously along a random direction, it becomes possible for the atom to change momentum family and to fall into a ground state characterized by a new value p' of the generalized momentum.

The time intervals between two successive spontaneous emissions by the same atom have a probability distribution $W_p(\tau)$ called the 'delay function' or the 'waiting time distribution' (see Section 2.3.3). In general, $W_p(\tau)$ is the squared

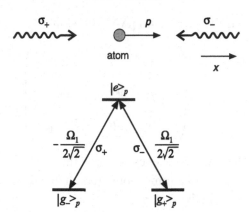

Fig. A.1. System of three atomic states (Λ system) coupled by two counterpropagating lasers, respectively σ_+ and σ_- circularly polarized. The coupling matrix elements corresponding to a $J = 1 \rightarrow J = 1$ transition are indicated. Note the opposite signs. The family $\mathcal{F}_p = \{|g_-\rangle_p, |g_+\rangle_p, |e\rangle_p\}$ characterized by the generalized momentum p along x is stable under the atom–laser coupling (absorption and stimulated emission), but the generalized momentum p (and therefore the family) changes when a fluorescence photon is emitted.

modulus of a sum of complex exponentials. However, under frequently valid approximations presented below, $W_p(\tau)$ reduces to a single (real) exponential law[1] thus allowing one to introduce a fluorescence rate $R(p)$, which is also the jump rate for the momentum (see eq. (2.6)).

In VSCPT, the rate of fluorescence $R(p)$ from ground states depends on p, and vanishes around $p = 0$. The momentum p therefore follows an inhomogeneous random walk as considered in Section 2.2, leading to subrecoil cooling.

The goal of this section is to explicitly calculate the fluorescence rate $R(p)$ as a function of the atom and laser parameters, in order to be able to derive the statistical quantities of interest in the following sections.

A.1.1.1 Effective Hamiltonian

In order to find the fluorescence rate $R(p)$ of an atom in the family $\mathcal{F}_p$, we introduce a non-Hermitian effective Hamiltonian which, in the basis

---

[1] Or, occasionally, to a sum of two (real) exponentials (see eq. (A.35)).

$\{|g_-\rangle_p, |g_+\rangle_p, |e\rangle_p\}$, has the expression [AAK89, Coh90]

$$\hat{H}_p = \hbar \begin{pmatrix} -\frac{kp}{M} & 0 & -\frac{\Omega_1}{2\sqrt{2}} \\ 0 & \frac{kp}{M} & \frac{\Omega_1}{2\sqrt{2}} \\ -\frac{\Omega_1}{2\sqrt{2}} & \frac{\Omega_1}{2\sqrt{2}} & -\tilde{\delta} - i\frac{\Gamma}{2} \end{pmatrix}. \tag{A.1}$$

The two lasers have the same frequency ω_L, and their wave-vectors are $\pm k$ along x. The atom has a mass M, an atomic Bohr frequency ω_A, and the lifetime of the upper level is Γ^{-1}. The effective detuning $\tilde{\delta}$,

$$\tilde{\delta} = \omega_L - \omega_A + \Omega_R, \tag{A.2}$$

takes into account the recoil frequency Ω_R

$$\Omega_R = \frac{\hbar k^2}{2M}. \tag{A.3}$$

The two lasers have the same intensity I, and the atom–laser coupling matrix elements have $1/\sqrt{2}$ factors because we define $\Omega_1 = \Gamma\sqrt{I/(2I_{sat})}$ as the Rabi frequency that would be associated with a two-level atom having the same radiative decay constant Γ of the excited state. In the case of metastable helium, $\lambda_A = 2\pi c/\omega_A = 2\pi/k = 1.0830$ μm, $\Gamma^{-1} = 100$ ns, $I_{sat} = 0.16$ mW/cm^2, and $\Omega_R = \Gamma/37.44$.

The diagonal terms $\pm kp/M$, which can be interpreted as Doppler shifts, come from the expansion of the matrix elements of the kinetic energy Hamiltonian (in eq. (A.1), we have dropped the diagonal term $\left(p^2/2M + \hbar^2 k^2/2M\right)$ times the unity matrix, which gives a uniform energy shift in the $\mathcal{F}_p$ family). The imaginary term i$\Gamma/2$ represents the radiative decay of the excited state. Because of the laser coupling between the excited state and the ground states, this imaginary term gives rise to an instability of the ground states that we interpret as the fluorescence rate, and that we want to calculate.

In order to evaluate the fluorescence rate of an atom in the ground subspace $\mathcal{G}_p = \{|g_-\rangle_p, |g_+\rangle_p\}$, we first diagonalize $\hat{H}_p$: each eigenvector $|u_j\rangle$ is a principal decay mode, with a decay rate Γ_j equal to the imaginary part of the corresponding eigenvalue λ_j (times a -2 factor)

$$\Gamma_j = -2\,\text{Im}\{\lambda_j\}. \tag{A.4}$$

Note that, because of conservation of the trace, we have

$$\sum_j \Gamma_j = \Gamma. \tag{A.5}$$

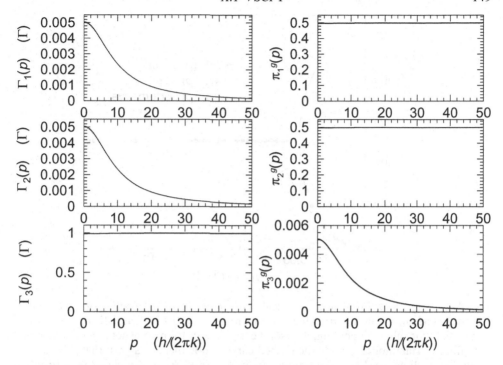

Fig. A.2. Decay constants $\Gamma_j(p)$, and corresponding statistical weights $\pi_j^g(p)$ in the ground subspace, at a gross scale (case of He*, $\Omega_1 = 0.1\Gamma$, $\tilde{\delta} = 0$). The solid curves result from an exact diagonalization of the effective Hamiltonian. The dotted curves in the plots of $\Gamma_1(p)$ and $\Gamma_2(p)$ exactly coincide with the solid curves at large p and are hardly distinguishable at small p. These dotted lines represent the result eq. (A.15) of the perturbative diagonalization at 'large' p values where there is no coherent population trapping ($kp/M \gg \Omega_1$, that is $p/\hbar k \gg \Omega_1/(2\Omega_R) \simeq 1.9$).

The statistical weight $\pi_j^g(p)$ associated with the decay rate Γ_j, for ground state atoms uniformly distributed among $|g_-\rangle_p$ and $|g_+\rangle_p$ (statistical mixture with equal weights 1/2), is given by the average

$$\pi_j^g(p) = \tfrac{1}{2} \left\| \langle g_- | u_j \rangle \right\|^2 + \tfrac{1}{2} \left\| \langle g_+ | u_j \rangle \right\|^2. \tag{A.6}$$

A.1.1.2 Exact diagonalization

Figure A.2 shows at a gross scale (p in the range [0, 50 $\hbar k$]) the result of the diagonalization of $\hat{H}_p$, plotted for a weak coupling ($\Omega_1 = 0.1\Gamma$). For each of the three eigenvalues, we have plotted, as a function of the momentum p, the decay rate Γ_j and the statistical weight $\pi_j^g(p)$, in the ground subspace, of the corresponding decay rate.

The decay rate Γ_3 is very close to Γ, and has a negligible statistical weight in the

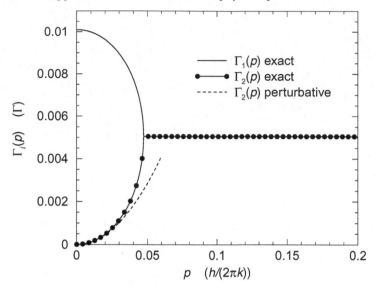

Fig. A.3. $\Gamma_1(p)$ and $\Gamma_2(p)$ at a finer scale near $p = 0$, for the same parameters as in Fig. A.2. The decay rate $\Gamma_2(p)$ exhibits a dip around $p = 0$, characteristic of VSCPT and responsible for subrecoil cooling. The solid curves result from an exact diagonalization of the effective Hamiltonian, while the dashed curve is obtained by a perturbative treatment around $p = 0$ (eq. (A.13)). The bifurcation at $p \simeq 0.047\hbar k$ is associated with lifting of the degeneracy between $\Gamma_1(p)$ and $\Gamma_2(p)$ occurring at small p.

ground subspace: it is obviously associated with the case of an atom mostly in the excited state, and is therefore irrelevant to our discussion that focuses on ground state atoms.

The decay rates Γ_1 and Γ_2 are almost equal at this scale, and they have the same statistical weight, close to $1/2$. They represent the fluorescence rate $R(p)$ in the ground subspace. They exhibit a Lorentzian variation with p that we interpret as due to the Doppler detuning.

Figure A.3 shows the same result at a finer scale near $p = 0$. In this region, the degeneracy between $\Gamma_2(p)$ and $\Gamma_1(p)$ is lifted. The jump rate $\Gamma_2(p)$ *tends quadratically to zero* with p: this is the main feature of VSCPT, that allows subrecoil cooling. Simultaneously, we see that $\Gamma_1(p)$ increases to $2R_0$, where R_0 is the fluorescence rate at the maximum of the gross scale Lorentzian. For atoms uniformly distributed in the ground subspace, the statistical weights $\pi_1^g(p)$ and $\pi_2^g(p)$ are both equal to $1/2$.

We are now going to obtain perturbative expressions of $\Gamma_1(p)$ and $\Gamma_2(p)$ in the two asymptotic situations of interest: close to $p = 0$ and at values of p beyond the bifurcation where there is no coherent population trapping. These analytic expressions will be used in the following sections to establish the connection with the simple models introduced in the main text.

A.1.1.3 Expansion around p = 0

Taking $p = 0$, we obtain the zeroth order (in p) eigenvalues and eigenvectors of $\hat{H}_p$. One eigenvalue, $\lambda_2(0)$ (labelled 2 in correspondence with the above discussion), is zero, and the corresponding decay constant is

$$\Gamma_2(0) = \Gamma_{NC}(0) = 0. \tag{A.7}$$

It is associated with the eigenvector

$$|u_2\rangle_0 = |\Psi_{NC}(0)\rangle = \frac{1}{\sqrt{2}}\left\{|g_-\rangle_0 + |g_+\rangle_0\right\} \tag{A.8}$$

which is the *trapping state*, not coupled to the excited state by the lasers, and introduced in the theoretical analysis of [AAK89]. To calculate the other eigenvalue, $\lambda_1(0)$, relevant to ground state atoms, we make the simplifying assumption[2] that:

$$\Omega_1 \ll \Gamma/2, \tag{A.9}$$

and we keep only the lowest order in Ω_1/Γ. Taking respectively the imaginary and real parts of $\lambda_1(0) = \delta_1'(0) - i\Gamma_1(0)/2$, we obtain the decay rate

$$\Gamma_1(0) = \Gamma_C(0) = \frac{\Omega_1^2}{\Gamma} \frac{1}{1 + (2\tilde{\delta}/\Gamma)^2} \tag{A.10}$$

and the energy shift (light shift)

$$\delta_1'(0) = \tilde{\delta}\,\frac{\Omega_1^2/\Gamma^2}{1 + (2\tilde{\delta}/\Gamma)^2}. \tag{A.11}$$

The corresponding eigenvector is

$$|u_1\rangle_0 = |\Psi_C(0)\rangle = \frac{1}{\sqrt{2}}\left\{|g_-\rangle_0 - |g_+\rangle_0\right\}. \tag{A.12}$$

We now consider the lowest-order (in p) non-zero term of $\Gamma_2(p)$ near $p = 0$:

$$\Gamma_2(p) = \Gamma_{NC}(p) \underset{p\to 0}{\simeq} \frac{4\Gamma}{\Omega_1^2}\left(\frac{kp}{M}\right)^2 = \frac{16\Gamma^2}{\Omega_1^2}\left(\frac{\Omega_R}{\Gamma}\right)^2\left(\frac{p}{\hbar k}\right)^2\Gamma. \tag{A.13}$$

The quadratic variation of eq. (A.13) around $p = 0$ is shown in Fig. A.3 (dashed curve) for $\Omega_1 = 0.1\Gamma$ and $\tilde{\delta} = 0$: it clearly coincides with the exact value

---

[2] This assumption allows us to present the essential features simply, but it plays no crucial role in the reasoning. Moreover, it is not difficult to generalize the calculations presented here when this assumption does not hold (see footnote 7 in this appendix (p. 158)). In that case, the quantities of interest depend on Ω_1 in a more complicated way.

obtained by the direct diagonalization of the whole matrix. A similar perturbative calculation in p gives:

$$\Gamma_1(p) = \Gamma_C(p) \underset{p \to 0}{\simeq} \Gamma_1(0) - \Gamma_2(p). \tag{A.14}$$

This relation also holds for the exact calculation presented above.

Remark. Note that the result of eq. (A.13) is independent of the detuning $\tilde{\delta}$. This can be interpreted (see [AAK89], Section 6.E) as compensation of the variation of the light shift of eq. (A.11) with $\tilde{\delta}$, by a similar variation of $\Gamma_C(0)$ (eq. (A.10)).

A.1.1.4 Behaviour out of the trapping dip

We now consider the case kp/M large compared to Ω_1. At zeroth order in Ω_1, the matrix $\hat{H}_p$ of eq. (A.1) is diagonal, and the only non-zero decay constant Γ_3 is associated with the excited state. To obtain the fluorescence rate of atoms in the ground state, we make a first-order (in Ω_1) perturbative diagonalization of $\hat{H}_p$. We find that the states $|g_-\rangle_p$ and $|g_+\rangle_p$ are associated with decay constants

$$\Gamma_{g_\pm}(p) = \frac{\Omega_1^2/8}{\left(\tilde{\delta} \pm kp/M\right)^2 + \Gamma^2/4}\, \Gamma. \tag{A.15}$$

The decay constants $\Gamma_{g_\pm}(p)$ exhibit a Lorentzian dependence with a half-width p_D at half-maximum, related to the Doppler effect, given by:

$$p_D = \frac{\Gamma M}{2k} = \frac{\Gamma}{4\Omega_R}\, \hbar k. \tag{A.16}$$

In the case $\tilde{\delta} = 0$, these two Lorentzians are identical and centred at the same value $p = 0$. To simplify the discussion and the equations, we consider only this case in the following[3]. For $p \ll p_D$, $\Gamma_{g_\pm}(p)$ does not depend on p and takes its maximum value

$$\Gamma_{g_\pm}(0) \simeq \frac{\Omega_1^2}{2\Gamma}. \tag{A.17}$$

For $p \gg p_D$, $\Gamma_{g_\pm}(p)$ has Lorentzian tails decaying as

$$\Gamma_{g_\pm}(p) = \frac{\Gamma}{8}\left(\frac{M\Omega_1}{kp}\right)^2 = \Gamma_{g_\pm}(0)\left(\frac{p_D}{p}\right)^2. \tag{A.18}$$

The two rates given in eq. (A.15) are plotted in Fig. A.2 (dotted line) for $\Omega_1 = 0.1\Gamma$ and $\tilde{\delta} = 0$. They cannot be distinguished from the exact eigenvalues obtained by diagonalization, for values of p beyond the bifurcation. We note in particular

---

[3] The generalization to $\tilde{\delta} \neq 0$ is not very difficult. Note in particular that eqs. (A.19a) and (A.19b) must be supplemented with $\tilde{\delta} \to \tilde{\delta}^{(F)} = F\tilde{\delta}$.

the good agreement even in the region where p is less than p_D (central region of the gross scale Lorentzian) but where $kp/M \gg \Omega_1$.

A.1.1.5 Case of a negligible Doppler effect

In this subsection, we show how a simple modification of the atomic and laser parameters, in one-dimensional σ_+/σ_- VSCPT quantum calculations, leads to fluorescence rates identical to the previous ones near $p = 0$ (VSCPT region), but with negligible Doppler decrease at large p. This will allow us to make the connection with the *unconfined model* introduced in the main text.

To make the Doppler effect negligible at large p, it is possible to shift the scale p_D to arbitrarily large values, while keeping unaffected the dip of the fluorescence rate around $p = 0$. For that purpose, we introduce a fictitious situation deduced from the real one considered above, by the transformation

$$\Gamma \longrightarrow \Gamma^{(F)} = F\Gamma, \tag{A.19a}$$

$$\Omega_1 \longrightarrow \Omega_1^{(F)} = \sqrt{F}\,\Omega_1, \tag{A.19b}$$

where F is a large enough factor. It is then clear in eqs. (A.13) and (A.17) that the fluorescence rate $\Gamma_{NC}(p)$ associated with the trapping state around $p = 0$ is unchanged, as well as the fluorescence rate $\Gamma_{g\pm}(0)$ at the maximum of the Lorentzian[4]. On the other hand, the half-width p_D of the Lorentzian is increased by the factor F

$$p_D \longrightarrow p_D^{(F)} = F\,p_D. \tag{A.20}$$

Thus, the Lorentzian decay is relegated to large values of p. By taking p_D larger than the largest momentum reached by the atoms during the studied interaction time θ, we simulate the unconfined model while keeping all the quantum equations of the problem. This is used in the quantum jump calculations of Section 8.3.2. The validity of this method is confirmed by independent calculations presented in [AlK96] (see Section 8.3.2 and in particular footnote 5, p. 111).

Figure A.4 shows the result of the exact diagonalization of the Hamiltonian $\hat{H}_p$ (eq. (A.1)) where we have made the transformation of eq. (A.19) with $F = 316$, the value chosen for the numerical simulations presented in Chapter 8. It confirms that the transformation (A.19) leaves the fluorescence rates $\Gamma_2(p)$ and $\Gamma_{g\pm}(p)$ unchanged[5] in the trapping dip and at small values of p while modifying it out of the

---

[4] The fluorescence rates remain the same in units of s^{-1}. Therefore time scales (in s) are left unchanged by the transformation. However, when specified in units of Γ^{-1}, times must be modified, according to: $t/\Gamma^{-1} \to (t/\Gamma^{-1})^{(F)} = F(t/\Gamma^{-1})$.

[5] The maximum value of the fluorescence rate is slightly (1%) larger in the real situation ($F = 1$) than in the fictitious situation ($F = 316$). This is because $\Omega_1^{(F)}/\Gamma^{(F)}$ is smaller than Ω_1/Γ by a factor $\sqrt{F}$. This factor improves the validity of the perturbative assumption (A.9) and therefore slightly modifies the value of the fluorescence rates. The larger the value of F, the better the approximation (A.9). The agreement would be perfect if the saturation terms were taken into account (see footnote 2, p. 151).

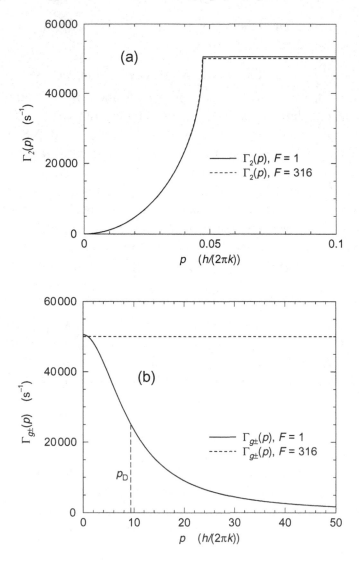

Fig. A.4. Exact fluorescence rate associated with the transformation (A.19). The solid curves represent the real situation ($F = 1$) with $\Omega_1 = 0.1\Gamma$, $\tilde{\delta} = 0$ and other parameters as in Fig. A.2 and Fig. A.3. The dashed curves represent the fictitious situation resulting from the transformation (A.19) with $F = 316$. Note the vertical scales in s^{-1} and not in Γ as in previous figures, because $\Gamma^{(F)}$ depends on F through (A.19). (a) In the vicinity of $p = 0$, $\Gamma_2(p)$ is unchanged by the transformation (A.19). (b) For large p, the large value of F pushes the Lorentzian decrease of the fluorescence rates $\Gamma_{g\pm}(p)$ towards larger values of p, out of the frame of this figure.

trapping dip, so that the size p_D of the region where $\Gamma_{g\pm}(p)$ remains approximately constant is extended by the factor F, from $p_D \simeq 9.4\hbar k$ to $p_D^{(F)} \simeq 3000\hbar k$.

A.1.2 Parameters of the random walk models

Inside the trap and in its vicinity, the three random walk models introduced in Chapter 3 (Section 3.2) coincide. They are all characterized by a quadratic jump rate around $p = 0$, and by a plateau. We first express the parameters τ_0 and p_0 characterizing this quadratic rate and the plateau as a function of the atom and laser parameters for one-dimensional σ_+/σ_- VSCPT. We then determine the parameter p_D beyond which the Doppler model is different from the unconfined model. We also justify the introduction of confining walls to mimic friction, and we give an estimate for p_{max}. Finally, we give the length Δp of the elementary step of the random walk in the momentum space.

A.1.2.1 Trapping region and plateau: p_0 and τ_0

Around $p = 0$ and in the neighbouring region, the three models have been defined by eqs. (3.5), (3.7), (3.8) and (3.9). These equations reveal the following features:

- a plateau with a constant fluorescence rate:

$$R(p) = \frac{1}{\tau_0} \quad \text{for} \quad p \ge p_0; \tag{A.21}$$

- a dip with a quadratic variation around $p = 0$:

$$R(p) = \frac{1}{\tau_0} \left(\frac{p}{p_0}\right)^2 \quad \text{for} \quad p \le p_0. \tag{A.22}$$

To make the connection with the quantum results, we first equal the constant rate of the plateau (eq. (A.21)) to the maximum rate $\Gamma_{g_\pm}(0)$ of the quantum jump rate out of the dip (eq. (A.17)). This gives the value of τ_0 as a function of the laser and atom parameters:

$$\tau_0 = 2\left(\frac{\Gamma}{\Omega_1}\right)^2 \Gamma^{-1}. \tag{A.23}$$

To make the connection in the trapping region, let us recall that according to the above analysis, an atom falling into a ground state in the region close to $p = 0$ can have two different evolutions. It can, with a probability $\pi_1 \simeq 1/2$, continue to fluoresce at a rate (see eqs. (A.10) and (A.14) for $\tilde{\delta} = 0$)

$$\Gamma_1(p) = \Gamma_C(p) \simeq \Gamma_C(0) = \frac{2}{\tau_0}. \tag{A.24}$$

This situation does not correspond to a trapping event since the random walk in the momentum space continues with a relatively large rate. But the atom falling into a ground state close to $p = 0$ can also, with a probability $\pi_2 \simeq 1/2$, fluoresce at a much smaller rate (see eq. (A.13))

$$\Gamma_2(p) \simeq \Gamma_{\text{NC}}(p) = \frac{4\Gamma}{\Omega_1^2} \left(\frac{kp}{M} \right)^2$$

corresponding to a trapped state $|\Psi_{\text{NC}}(p)\rangle$. The identification with the jump rate given by eq. (A.22) leads to:

$$\frac{1}{\tau_0\, p_0^2} = \frac{4\Gamma}{\Omega_1^2} \left(\frac{k}{M} \right)^2 \qquad\qquad\qquad (\text{A.25})$$

or, in reduced units:

$$(\Gamma\tau_0) \left(\frac{p_0}{\hbar k} \right)^2 = \frac{1}{16} \left(\frac{\Omega_1}{\Omega_\text{R}} \right)^2 . \qquad\qquad (\text{A.26})$$

We thus find

$$p_0 = \frac{1}{2\sqrt{2}} \frac{M\,\Omega_1^2}{k\Gamma} \qquad\qquad\qquad\qquad (\text{A.27})$$

or, in reduced units,

$$\frac{p_0}{\hbar k} = \frac{1}{2^{5/2}} \frac{\Omega_1^2}{\Omega_\text{R}\Gamma} . \qquad\qquad\qquad (\text{A.28})$$

A.1.2.2 Dependence on laser intensity

The jump rate $R(p)$ resulting from eq. (A.21) and eq. (A.22) together with eq. (A.23) and eq. (A.28) is represented in Fig. A.5 for two values of Ω_1. The dependence on laser intensity, which is proportional to Ω_1^2, is worth noting. In the plateau outside the trapping dip, the jump rate $R(p)$ increases with the intensity. This agrees with intuition: a larger laser intensity corresponds to a larger rate of fluorescence. But, inside the trapping dip, $R(p) = \Gamma_2(p)$ *decreases* with Ω_1[6]. This behaviour can actually be explained, invoking the instability in the ground states due to the coupling to the excited state (see [AAK89], Section 3.C).

The existence of opposite dependences of $R(p)$ on laser intensity inside and outside the trapping dip plays a crucial role in the cooling optimization (see Section 9.3).

---

[6] On the contrary, $\Gamma_1(p)$ increases with Ω_1, as intuition suggests.

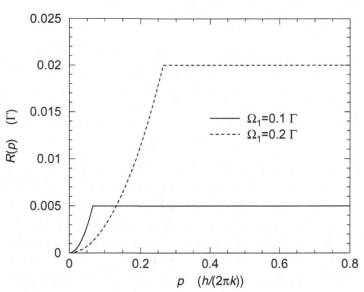

Fig. A.5. Model of the jump rate for VSCPT at two different laser intensities. Increasing the laser intensity I, i.e. increasing Ω_1 which scales as $I^{1/2}$, implies: (i) increasing the size p_0 of the trapping dip; (ii) increasing the jump rate $1/\tau_0$ out of the trapping dip; (iii) decreasing the jump rate in the trapping dip.

A.1.2.3 Doppler tail: p_D

The Doppler recycling model must coincide with the quantum calculations at large values of p. Except at very small values of p, the quantum calculations predict a Lorentzian behaviour of the fluorescence rate centred at $p = 0$ (eq. (A.15) for $\tilde{\delta} = 0$). The model approximates this Lorentzian variation by:

- a plateau (already introduced above) where the constant rate is equal to the maximum of the quantum Lorentzian (cf. eq. (A.23)) for $p \leq p_D$:

$$R(p) = \frac{1}{\tau_0} = \Gamma_{g\pm}(0) = \Gamma\frac{\Omega_1^2}{2\Gamma}; \tag{A.29}$$

- a wing coinciding with the Lorentzian tail given by the quantum calculation at large $p \geq p_D$ (cf. eq. (A.18)):

$$R(p) = \frac{1}{\tau_0}\left(\frac{p_D}{p}\right)^2 = \frac{\Gamma}{8}\left(\frac{M\Omega_1}{kp}\right)^2 = \frac{\Gamma}{32}\left(\frac{\Gamma}{\Omega_R}\right)^2\left(\frac{\Omega_1}{\Gamma}\right)^2\left(\frac{\hbar k}{p}\right)^2. \tag{A.30}$$

The parameter p_D is thus taken equal to:

$$p_D = \frac{\Gamma M}{2k} \tag{A.31}$$

or, equivalently in reduced units:

$$\frac{p_D}{\hbar k} = \frac{1}{4}\frac{\Gamma}{\Omega_R}. \tag{A.32}$$

A.1.2.4 Discussion: comparison between quantum calculations and statistical models

Figure A.6 shows a comparison between the *Doppler model* and the results of the exact quantum calculations for one-dimensional σ_+/σ_- VSCPT. The model coincides with the exact results[7] in two extreme situations: around $p = 0$ and at very large p. *This is exactly what is required to describe correctly the asymptotic limits of very long trapping times on the one hand* $(p \to 0)$ *and of very long recycling times on the other* $(p \to \infty)$. The intermediate region where the model significantly differs from the exact calculation plays no role in the evaluation of the tails of the broad distributions of the trapping and recycling times: in particular, the value $R = 1/\tau_0$ of the height of the plateau is irrelevant since the relevant quantities are $\tau_0 p_0^2$ on the one hand and p_0^2/τ_0 on the other, both quantities being fully determined by the behaviour of the jump rate around $p = 0$, and for very large p. We therefore expect the model to reproduce very well the results of the quantum calculations for the asymptotic limit of long interaction times, for quantities that depend only on the tails of the trapping and recycling times distributions.

Figure A.7 shows a comparison between the *unconfined model*, and the results of the exact quantum calculations for one-dimensional σ_+/σ_- VSCPT, with modified parameters (case of negligible Doppler effect, see Section A.1.1.5). The agreement is excellent around $p = 0$ (long trapping times) and in the plateau (long recycling times). We therefore here also expect the model to reproduce very well the results of the quantum calculations for the asymptotic limit of very long interaction times, for quantities that depend only of the tails of the trapping and recycling times distributions. For the unconfined model, the relevant quantities for the correspondence are $\tau_0 p_0^2$, which characterizes the trapping region, and $1/\tau_0$, which is the jump rate in the plateau. Notice that the perturbative (in Ω_1) expressions of $\tau_0 p_0^2$ (eq. (A.25)) and of τ_0 (eq. (A.23)) have a much wider range of validity, since the transformation (A.19) multiplies the ratios Ω_1/Γ by a factor $F^{-1/2}$ much smaller than one, so that the perturbative parameter is multiplied by the same small factor (see footnote 5, p. 153).

---

[7] When we are not in the perturbative limit $\Omega_1 \ll \Gamma$, the agreement is not as good around $p = 0$, but one may easily generalize eq. (A.13) by using a more exact expression (in Ω_1) of the quadratic (in p) jump rate around $p = 0$, rather than the expression (A.13) of Section A.1.1.3. Notice that, by contrast, the perturbative (in Ω_1) expression (A.30) of $R(p)$ at large p values remains valid for large values of p ($kp/M \gg \Omega_1$).

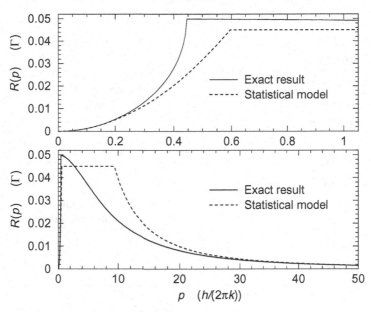

Fig. A.6. Comparison between the Doppler model (dashed curve), and the exact quantum calculation (solid curve). Parameters as in figure A.2 and figure A.3, except for $\Omega_1 = 0.3\Gamma$ ($\tilde{\delta} = 0$). The agreement is excellent around $p = 0$ (limit of long trapping times) and for large p values (limit of long recycling times). This is why the model describes the long trapping times and the long recycling times correctly. The plateau of the statistical model is arbitrarily defined as the maximum of the quantum optics result calculated perturbatively ($\Omega_1 \ll \Gamma$), which is not valid here ($\Omega_1 = 0.3\Gamma$).

A.1.2.5 Confining walls: p_{max}

As explained in Section 8.5 and in the introduction to this appendix, there are several situations (in one, two or three dimensions) in which a friction force appears out of the trapping region. This friction force acts as a confining force in the momentum space, which pushes the atoms back towards $p = 0$. It prevents the atomic random walk in momentum space from reaching large values of p, and we mimic its effect by introducing confining 'walls' at $p = p_{\mathrm{max}}$ in momentum space. To obtain a reasonable value for p_{max}, we recall that optimized Sisyphus cooling leads to a steady-state momentum distribution with a rms half-width of a few $\hbar k$. We will therefore take a value of a few $\hbar k$, typically:

$$p_{\mathrm{max}} \simeq 3\hbar k. \tag{A.33}$$

Notice however that this value is chosen somewhat arbitrarily. It can be fixed more precisely in each specific cooling configuration, using the theory of Sisyphus cooling [DaC89, UWR89, CaM95].

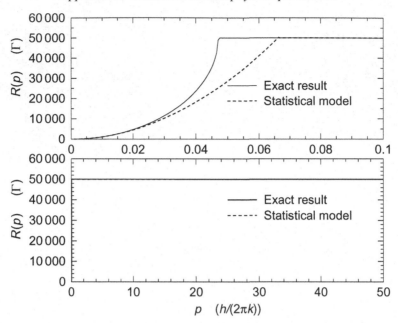

Fig. A.7. Comparison between the unconfined model (dashed line), and the modified quantum calculation without Doppler effect (solid line). Same parameters as in Fig. A.4 ($\Omega_1 = 0.1\Gamma$; $\tilde{\delta} = 0\Gamma$; $F = 316$). The agreement is excellent around $p = 0$ (limit of long trapping times) and for large p values (limit of long recycling times). On the lower graph, the two curves are not distinguishable. Note the scale of $R(p)$ in s^{-1} and not in Γ^{-1} in order to compare different F values.

A.1.2.6 Elementary step of the random walk: Δp

The rms length Δp of the elementary step of the random walk in momentum space is an important parameter of the models. For one-dimensional σ_+/σ_- VSCPT, each step comprises one absorption of a laser photon along x and one spontaneous scattering along a random direction in three dimensions, with an isotropic distribution[8].

The rms exchanged momentum along x is therefore

$$\Delta p = \sqrt{\frac{4}{3}}\, \hbar k. \tag{A.34}$$

For other one-dimensional laser configurations or for two- and three-dimensional cooling, specific calculations of Δp must be made.

---

[8] In a more realistic model, one should take into account the dipolar pattern of spontaneous emission. This would give $\Delta p = \sqrt{7/5}\hbar k \simeq 1.183\hbar k$, which is close to the result of eq. (A.34) ($\sqrt{4/3} \simeq 1.155$). See Section IV.3.4.1 in [Bar95].

A.1.3 Trapping time distribution: τ_b

We have established the connection between the atomic parameters and the quantity $\tau_0 p_0^2$ that completely characterizes the quadratic fluorescence rate of the trapped atoms around $p = 0$. We can therefore deduce the parameter τ_b appearing in the statistical distribution of the trapping times which is the same for the three recycling statistical models.

We use eq. (3.23) relative to an exponential law for the waiting times, given a certain jump rate. We recall (see Section A.1.2.1) that an atom entering the trapping region at momentum p has a probability $\pi_2 \simeq \frac{1}{2}$ of falling into a trapping state[9], and therefore having a jump rate following eq. (A.13), and a probability $\pi_1 \simeq \frac{1}{2}$ of falling into a non-trapping state, and therefore having a jump rate $\Gamma_1(p) \simeq \Gamma_C(0)$ following eq. (A.10). The conditional trapping time distribution[10] $P(\tau|p) = W_p(\tau)$ for such an atom thus reads:

$$P(\tau|p) = \tfrac{1}{2}\Gamma_1(0)e^{-\Gamma_1(0)\tau} + \tfrac{1}{2}\Gamma_2(p)e^{-\Gamma_2(p)\tau}. \tag{A.35}$$

The total probability distribution $P(\tau)$ (cf. eq. (3.20)) is therefore

$$
\begin{aligned}
P(\tau) &= \int_{-p_{\text{trap}}}^{p_{\text{trap}}} \rho(p_x) P(\tau|p_x)\, \mathrm{d}p_x \\
&= \frac{1}{2}\Gamma_1(0)e^{-\Gamma_1(0)\tau} + \frac{1}{2}\int_{-p_{\text{trap}}}^{p_{\text{trap}}} \rho(p_x)\Gamma_2(p_x)e^{-\Gamma_2(p_x)\tau}\, \mathrm{d}p_x.
\end{aligned} \tag{A.36}
$$

At large τ, the first term becomes exponentially negligible compared to the second, which vanishes only as a power law in τ. This second term is equal to one half the distribution calculated in eq. (3.23):

$$P(\tau) \underset{\tau\to\infty}{\simeq} \frac{1}{2} \times \frac{\Gamma(1/2)}{2} \frac{\tau_{\text{trap}}^{1/2}}{2\,\tau^{3/2}}. \tag{A.37}$$

Using the definition of eq. (3.32) ($P(\tau) = \tau_b^{1/2}/2\tau^{3/2}$) for τ_b and eq. (3.15) for τ_{trap}, we first obtain

$$\tau_b = \frac{\pi}{16}\left(\frac{p_0}{p_{\text{trap}}}\right)^2 \tau_0, \tag{A.38}$$

which can then be written, using eq. (A.26):

$$\tau_b = \frac{\pi}{2^8}\left(\frac{\Omega_1}{\Omega_R}\right)^2 \left(\frac{\hbar k}{p_{\text{trap}}}\right)^2 \Gamma^{-1}. \tag{A.39}$$

---

[9] In the case where the statistical weight $\pi_2(p)$ would vary significantly with p around $p = 0$, this variation could modify the exponent μ of the power-law characterizing the trapping times.

[10] This is the only case in this work where the delay function (see Sections 2.3.3 and 2.4.2) is not reduced to a single exponential.

Notice that the arbitrary parameter p_{trap} should disappear at the end of calculations of measurable quantities. The same remark holds for the analogous quantities derived below.

This result can be generalized to the case of D dimensions. Using eq. (3.32), and introducing the same factor $1/2$ as above, we can generalize eq. (A.38) to

$$\tau_{\text{b}} = \left(\frac{D}{4} \Gamma(D/2) \right)^{2/D} \left(\frac{p_0}{p_{\text{trap}}} \right)^2 \tau_0, \tag{A.40}$$

and eq. (A.39) to

$$\tau_{\text{b}} = \frac{1}{16} \left(\frac{D}{4} \Gamma(D/2) \right)^{2/D} \left(\frac{\Omega_1}{\Omega_{\text{R}}} \right)^2 \left(\frac{\hbar k}{p_{\text{trap}}} \right)^2 \Gamma^{-1}. \tag{A.41}$$

Note, however, that choosing the exact value of Ω_1 would require further work. If $D > 1$, the amplitude of the total laser field varies in space and its average must be carefully evaluated.

The expression (A.41) is valid in any dimension. However, if $D > \alpha$, the average value $\langle \tau \rangle$ is finite and the trapping time distribution $P(\tau)$ is then more conveniently characterized by $\langle \tau \rangle$. Taking into account both coupled states and non-coupled states (and using eqs. (A.14) and (A.23))), one obtains:

$$\langle \tau \rangle \simeq \tau_0 = 2 \left(\frac{\Gamma}{\Omega_1} \right)^2 \Gamma^{-1}. \tag{A.42}$$

A.1.4 Recycling time distribution

In this section we want to express the parameters characterizing the recycling time distribution as a function of the atom and laser parameters. Since the three statistical recycling models considered in this work differ in the recycling region out of the trap, we will have to consider the three cases separately.

A.1.4.1 Doppler model: $\hat{\tau}_{\text{b}}$

For the Doppler model, in one dimension, the probability distribution of the recycling times reads $\hat{P}(\hat{\tau}) = \hat{\tau}_{\text{b}}^{1/4}/(4\hat{\tau}^{5/4})$, according to eq. (3.49), where the typical time $\hat{\tau}_{\text{b}}$ is given by eq. (3.50) (with the numerical factor of eq. (B.23))

$$\hat{\tau}_{\text{b}} = (0.3296\ldots)^4 \frac{\Delta p^6}{p_{\text{trap}}^4 p_{\text{D}}^2} \tau_0. \tag{A.43}$$

It is completely characterized by $\tau_0\, p_D^{-2}$ and Δp. Using eq. (A.34) for Δp, eq. (A.32) for p_D and eq. (A.23) for τ_0, we obtain

$$\hat{\tau}_b = \frac{(0.3296\ldots)^4 2^{11}}{3^3} \left(\frac{\hbar k}{p_{\text{trap}}}\right)^4 \left(\frac{\Omega_R}{\Omega_1}\right)^2 \Gamma^{-1} = 0.895\ldots \left(\frac{\hbar k}{p_{\text{trap}}}\right)^4 \left(\frac{\Omega_R}{\Omega_1}\right)^2 \Gamma^{-1}.$$

(A.44)

This result is valid only in one dimension.

A.1.4.2 Unconfined model: $\hat{\tau}_b$

The recycling time distribution of the unconfined model corresponds to a standard homogeneous random walk (Section 3.4.2), with a constant jump rate $1/\tau_0$. According to eq. (3.46), we have

$$\hat{P}(\hat{\tau}) = \frac{\hat{\tau}_b^{1/2}}{2\,\hat{\tau}^{3/2}}$$

(A.45)

with

$$\hat{\tau}_b = \frac{1}{2\pi} \left(\frac{\Delta p}{p_{\text{trap}}}\right)^2 \tau_0,$$

(A.46)

which gives, using eq. (A.23) and eq. (A.34),

$$\hat{\tau}_b = \frac{4}{3\pi} \left(\frac{\Gamma}{\Omega_1}\right)^2 \left(\frac{\hbar k}{p_{\text{trap}}}\right)^2 \Gamma^{-1}.$$

(A.47)

Again, this result is valid only in one dimension.

A.1.4.3 Confined model: $\langle\hat{\tau}\rangle$

In the confined model, the homogeneous random walk out of the trap is confined by reflecting walls at $p = p_{\text{max}}$. As explained in Section 3.4.4, the average first return time $\langle\hat{\tau}\rangle$ is finite in this case. It is given by eq. (3.56):

$$\langle\hat{\tau}\rangle = \tau_0 \left(\frac{p_{\text{max}}}{p_{\text{trap}}}\right)^D,$$

which, using eq. (A.23), can also be written

$$\langle\hat{\tau}\rangle = 2 \left(\frac{p_{\text{max}}}{p_{\text{trap}}}\right)^D \left(\frac{\Gamma}{\Omega_1}\right)^2 \Gamma^{-1}.$$

(A.48)

The position p_{max} of the confining walls is given in eq. (A.33). Recall, however, that p_{max} has been defined somewhat arbitrarily, and therefore results depending explicitly on p_{max} should be considered with some caution.

A.2 Raman cooling

In this section, we first present a model jump rate (Section A.2.1) for Raman cooling, from which we infer the parameters of the random walk models (Section A.2.2). The parameters of the trapping time distribution (Section A.2.3) and of the recycling time distribution (Section A.2.4) are then derived.

A.2.1 Jump rate

In Raman cooling [KaC92], the jump rate $R(p)$ results from the superposition of individual jump probabilities $P_i(p)$ created by a sequence of light pulses, labelled by i. The jump probability $P_i(p)$ created by a single light pulse is given, in the limit of a weak excitation, by the squared Fourier transform of the time evolution of an effective Rabi frequency $\Omega_{1,\,\mathrm{eff}}(t)$[11]. Thus, by changing the time dependence of the intensity (proportional to the square of the Rabi frequency) of the light pulse, one can obtain a variety of jump probabilities $P_i(p)$. For each pulse i, the shape, the central position and the magnitude of the jump probabilities $P_i(p)$ can all be varied. This gives a unique flexibility with which to tailor the jump rates $R(p)$ achievable in Raman cooling.

For the simplicity of the theoretical analysis, we consider here a pulse sequence based on square time Raman pulses, i.e. where $\Omega_{1,\,\mathrm{eff}}(t)$ is a square pulse. This sequence has been shown, in a one-dimensional experiment, to be both efficient and simple to implement in practice [RBB95]. It leads to an exponent $\alpha = 2$. Other sequences, such as those based on Blackman pulses [KaC92] ($\alpha \simeq$ 3–4 [RBB95, Rei96]), could be considered as well.

Let us first describe the jump probability $P_i(p_x)$ created by a *single pulse* of duration $\tau_{\mathrm{p},i}$ (more precisely, $P_i(p_x)$ is the probability that the atom will emit a spontaneous photon after one pulse). The laser beams are along the x axis so that the light pulse acts only on the x component of the atomic momentum **p**. When the light intensity of a pulse of duration $\tau_{\mathrm{p},i}$ is adapted so that the maximum of this probability is maximized to one ('π pulse'), we have[12]

$$P_i(p_x) = \left\{ \sin\left[\frac{\tau_{\mathrm{p},i}}{2}\left(\tilde{\delta}_i - \frac{2kp_x}{M} \right) \right] \Big/ \left[\frac{\tau_{\mathrm{p},i}}{2}\left(\tilde{\delta}_i - \frac{2kp_x}{M} \right) \right] \right\}^2, \qquad (A.49)$$

where $\tilde{\delta}_i$ is the detuning between the frequency difference of the lasers and the Raman transition (taking into account the atomic recoil terms), k is the wave-vector

---

[11] For the physics of these lights pulses, we refer the reader to the original article [KaC92] and, for a more detailed treatment, to [Rei96].

[12] This expression of the jump probability of a single pulse is the simplest, but not the most accurate, one can derive. Better approximations can be found in [Rei96]. They only introduce minor quantitative differences in the prefactors, which are irrelevant for our purposes here.

of the lasers acting on the Raman transition and M is the atomic mass. As we do not want to excite atoms in $p_x = 0$, we fix the detuning $\tilde{\delta}_i$ so that the first cancellation of $P_i(p_x)$ to the left[13] of its maximum occurs in $p_x = 0$:

$$\tilde{\delta}_i = \frac{2\pi}{\tau_{p,i}}. \tag{A.50}$$

To summarize, the single pulse jump probability (see Fig. A.8) is

$$P_i(p_x) = \left[\sin\left(\pi - \frac{k\tau_{p,i}}{M}p_x\right) \Big/ \left(\pi - \frac{k\tau_{p,i}}{M}p_x\right)\right]^2. \tag{A.51}$$

For a given atom, i.e. a given k and a given M, it depends on a single adjustable parameter, $\tau_{p,i}$. It presents a maximum in p_i:

$$p_i = \frac{\pi M}{k\tau_{p,i}} = \frac{\pi}{2}\frac{1}{\Omega_R \tau_{p,i}}\hbar k, \tag{A.52}$$

where the recoil frequency Ω_R is defined, as in eq. (A.3), by

$$\Omega_R = \frac{\hbar k^2}{2M}.$$

The probability $P_i(p_x)$ cancels at

$$p_x = \ldots, \ -2p_i, \ -p_i, \ 0, \ 2p_i, \ 3p_i, \ldots. \tag{A.53}$$

In the vicinity of $p_x = 0$, one has:

$$P_i(p_x) \xrightarrow[p_x \to 0]{} \left(\frac{k\tau_{p,i}}{\pi M}p_x\right)^2 = \left(\frac{p_x}{p_i}\right)^2 = \frac{4}{\pi^2}\left(\Omega_R\tau_{p,i}\right)^2\left(\frac{p_x}{\hbar k}\right)^2. \tag{A.54}$$

To achieve an efficient cooling in one dimension, a *sequence of pulses* (see Fig. A.9) is needed to excite all atomic momenta p_x except in the vicinity of $p_x = 0$. Interestingly, this can be realized for $p_x > 0$ in a relatively small time by a 'geometric' sequence of pulses of durations

$$\tau_{p,1}, \ \tau_{p,2} = \frac{\tau_{p,1}}{2}, \ \ldots, \ \tau_{p,i} = \frac{\tau_{p,1}}{2^{i-1}}, \ \ldots, \tag{A.55}$$

centred at

$$p_1, \ p_2 = 2p_1, \ \ldots, \ p_i = 2^{i-1}p_1, \ \ldots, \tag{A.56}$$

with the $\tau_{p,i}$'s and p_i's being related by eq. (A.52) and the detunings $\tilde{\delta}_i$'s following

---

[13] At this point, the cancellation could have been chosen to occur at the right of the maximum of $P_i(p_x)$. The sign of the position of the centre p_i of $P_i(p_x)$ (see eq. (A.52)) becomes important only when friction forces are taken into account (see below).

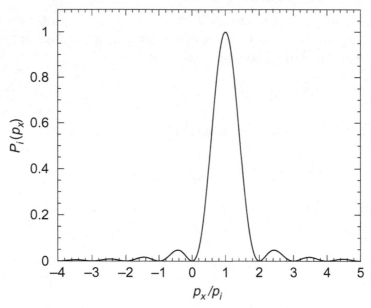

Fig. A.8. Excitation probability $P_i(p_x)$ generated by a single light pulse of Raman cooling.

eq. (A.50). To cover $p_x < 0$, we take a symmetric sequence of pulses, with durations

$$\tau_{\mathrm{p},-1} = \tau_{\mathrm{p},1}, \ \tau_{\mathrm{p},-2} = \tau_{\mathrm{p},2}, \ \ldots, \tag{A.57}$$

centred at

$$p_{-1} = -p_1, \ p_{-2} = -p_2, \ \ldots, \tag{A.58}$$

and with detunings

$$\tilde{\delta}_{-1} = -\tilde{\delta}_1, \ \tilde{\delta}_{-2} = -\tilde{\delta}_2, \ \ldots. \tag{A.59}$$

The detunings $\tilde{\delta}_i$ and intensities of the pulses are adapted so that each pulse creates a jump probability similar to eq. (A.51):

$$P_i(p_x) = \left[\frac{\sin\left(\pi - \mathrm{sign}(i)\frac{k\tau_{\mathrm{p},i}}{M}\,p_x\right)}{\pi - \mathrm{sign}(i)\frac{k\tau_{\mathrm{p},i}}{M}\,p_x} \right]^2. \tag{A.60}$$

Thus, the whole sequence of pulses preserves the dark state character of $p_x = 0$.

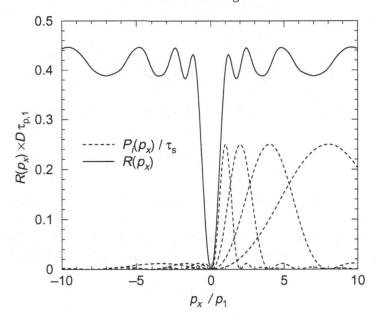

Fig. A.9. Jump rate $R(p_x)$ for the geometric sequence of square pulses. The jump rates $P_i(p_x)/\tau_s$ created by the first individual pulses, centred at p_1, $p_2 = 2p_1$, $p_3 = 4p_1, \ldots$, are represented by dashed curves. The pulses centred at $p_{-i} < 0$ are omitted for clarity. The resulting jump rate $R(p_x)$ is quadratic in the vicinity of $p_x = 0$ (eq. (A.66a)) and approximately constant for $|p_x| \gtrsim p_1$ (eq. (A.66b)).

The duration τ_s of the sequence[14] covering the whole p_x axis is

$$\tau_s = \sum_{i=1}^{\infty} \left(\tau_{p,-i} + \tau_{p,i} \right) = 2 \sum_{i=1}^{\infty} \frac{\tau_{p,1}}{2^{i-1}} = 4\tau_{p,1}. \qquad (A.61)$$

The main asset of the geometric sequence is to enable nearly uniform excitation of the whole p_x axis, except in the vicinity of $p_x = 0$, in a time τ_s which is only finitely larger than the durations $\tau_{p,\pm1}$ of the longest pulses.

The *jump rate* $R(p_x)$ resulting from the sequence is

$$R(p_x) = \frac{1}{\tau_s} \sum_{i=1}^{\infty} [P_{-i}(p_x) + P_i(p_x)], \qquad (A.62)$$

as shown in Fig. A.9. In the vicinity of $p_x = 0$, more precisely for $|p_x| \ll p_1$,

---

[14] Apart from the 'Raman' pulses presented here, Raman cooling requires auxiliary pulses, called 'repumping' pulses. These play only a minor role for our statistical analysis, even if they are essential in practice. Indeed, the durations of the 'repumping' pulses are negligible compared with the durations of the Raman pulses.

using eq. (A.54) and eq. (A.61), one has

$$R(p_x) \xrightarrow[p_x \to 0]{} \frac{2}{3\tau_{p,1}} \left(\frac{p_x}{p_1}\right)^2 = \frac{8}{3\pi^2} \Omega_R^2 \tau_{p,1} \left(\frac{p_x}{\hbar k}\right)^2. \qquad (A.63)$$

For large values of $|p_x|$, Fig. A.9 shows that $R(p_x)$ is approximately constant

$$R(p_x) \simeq \frac{5}{3\tau_s} = \frac{5}{12\tau_{p,1}}. \qquad (A.64)$$

Up to now, we have dealt only with Raman cooling in one dimension. The generalization of Raman cooling to several dimensions is obvious. One simply adds laser pulses along all the directions that are to be cooled. Thus, the duration of a pulse sequence in D dimensions is, after eq. (A.61) for one dimension,

$$\tau_s = 4D\tau_{p,1}. \qquad (A.65)$$

And the resulting jump rate $R(p)$ is

$$p = \|\mathbf{p}\| \ll p_1: \quad R(p) \xrightarrow[p \to 0]{} \frac{2}{3D\tau_{p,1}} \left(\frac{p}{p_1}\right)^2 = \frac{8}{3D\pi^2} \Omega_R^2 \tau_{p,1} \left(\frac{p}{\hbar k}\right)^2, \qquad (A.66a)$$

$$p \gtrsim p_1: \quad R(p) \simeq \frac{5}{12D\tau_{p,1}}. \qquad (A.66b)$$

Finally, we note that each pulse generates not only a probability $P_i(p_x)$ of spontaneous emission but also a deterministic momentum transfer of magnitude $2\hbar k$. The sign of the deterministic transfer can be chosen so that each pulse tends to reduce the magnitude of the atomic momenta on which it acts preferentially. Thus, pulses with $p_i > 0$ (respectively $p_i < 0$) induce transfers of $-2\hbar k$ (respectively $+2\hbar k$). These deterministic transfers provide a very efficient way of *confining* atomic momenta.

A.2.2 Parameters of the random walk models

Having calculated the jump rate $R(p)$, one can now derive the random walk model suitable for Raman cooling. It is clear from Fig. A.9 and from the previous section that the proposed sequence of pulses presents all the features of the *confined model* introduced in Section 3.2: a trapping dip with a power-law dependence of $R(p)$ around $p = 0$; a constant jump rate at large p; and confining forces.

A.2.2.1 Trapping region and plateau: p_0 and τ_0

We want to connect the jump rate computed for Raman cooling (cf. eq. (A.66a,b)) with the jump rate of the confined model (cf. eq. (3.5) and eq. (3.7)):

$$p \leq p_0: \quad R(p) = \frac{p^\alpha}{\tau_0 p_0^\alpha},$$

$$p \geq p_0: \quad R(p) = \frac{1}{\tau_0}.$$

This identification is possible with

$$\alpha = 2, \tag{A.67}$$

$$\tau_0 p_0^2 = \frac{3 D \pi^2 (\hbar k)^2}{8 \Omega_R^2 \tau_{p,1}}, \tag{A.68}$$

$$\tau_0 = \frac{12 D \tau_{p,1}}{5}, \tag{A.69}$$

and therefore, using eq. (A.52),

$$p_0 = \sqrt{\frac{5}{8}} \, p_1 = \sqrt{\frac{5}{12}} \, \frac{\pi}{\Omega_R \tau_{p,1}} \, \hbar k. \tag{A.70}$$

Figure A.10 shows that the rate of the statistical model and the realistic rate $R(p_x)$ of eq. (A.62) are close to each other.

As in the case of VSCPT (see p. 156), we observe that the dependences of $R(p)$ on the important cooling parameter, here $\tau_{p,1}$, are opposite inside and outside the trapping dip.

A.2.2.2 Confining walls: p_{max}

The deterministic momentum transfers associated with each Raman pulse provide a very efficient friction mechanism, confining the momenta to a few photon recoils. One typically has, as for VSCPT with friction (eq. (A.33)):

$$p_{max} \simeq 3 \hbar k. \tag{A.71}$$

A.2.2.3 Elementary step of the random walk: Δp

Each pulse i generates, with probability $P_i(p_x)$, a momentum transfer $-\text{sign}(p_i) 2 \hbar k$ and a random momentum transfer due to spontaneous emission along a random direction in three dimensions, with an isotropic distribution (see footnote 8, p. 160).

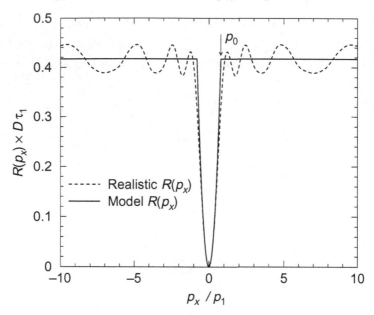

Fig. A.10. Comparison between the jump rate of the statistical model of Raman cooling and the realistic jump rate. The two jump rates agree perfectly in the trapping dip ($|p_x| \ll p_1$) and agree on average at large momenta ($|p_x| \gtrsim p_1$). This ensures that the statistical approach will be valid at long times θ.

The rms exchanged momentum along x is therefore[15]

$$\Delta p = \sqrt{\frac{13}{3}} \hbar k. \tag{A.72}$$

This result is valid in one dimension. Similar calculations can be made for two- and three-dimensional cases.

Here, we neglect the momentum transfers due to the repumping pulses which are specific to each experimental configuration. They can change the value of Δp by a factor of less than two.

A.2.3 Trapping time distribution: τ_b

An atom of momentum p_x submitted to a Raman pulse i undergoes a Bernoulli process: its momentum is either changed (probability $P_i(p_x)$) or it is unchanged (probability $1 - P_i(p_x)$). When $|p_x| \ll p_i$, as $P_i(p_x) \ll 1$, one can consider that the trapping time $\tau(p_x)$ resulting from the Bernoulli process is an exponential random variable with an average value $1/(P_i(p_x)/\tau_s)$. The trapping time distribution

---

[15] We give this value for completeness. However, it is not needed for the confined model, which is relevant for Raman cooling.

$P(\tau)$ is therefore (as in eq. (3.23) (exponential model))

$$P(\tau) \underset{\tau \gg \tau_{\text{trap}}}{=} \frac{\Gamma(1/2)\tau_{\text{trap}}^{1/2}}{4\tau^{3/2}}.$$

Using the definition of eq. (3.32) ($P(\tau) = \tau_{\text{b}}^{1/2}/(2\,\tau^{3/2})$) for τ_{b}, eq. (3.15) for τ_{trap} and eq. (A.68) for $\tau_0 p_0^2$, we obtain the correspondence

$$\tau_{\text{b}} = \frac{3D\pi^3}{32\Omega_R^2 \tau_{\text{p},1}} \left(\frac{\hbar k}{p_{\text{trap}}}\right)^2. \tag{A.73}$$

The expression (A.73) is valid in any dimension. However, if $D > \alpha$, the average value $\langle \tau \rangle$ is finite and the trapping time distribution $P(\tau)$ is then more conveniently characterized by $\langle \tau \rangle$. We then obtain, using eq. (3.35)[16],

$$\langle \tau \rangle = \frac{3D\pi^2}{4\Omega_R \tau_{\text{p},1}} \left(\frac{\hbar k}{p_{\text{trap}}}\right)^2. \tag{A.74}$$

A.2.4 Recycling time distribution: $\langle \hat{\tau} \rangle$

Being in the confined model, the average first return time $\langle \hat{\tau} \rangle$ is finite and is given by eq. (3.56). Using eq. (A.69) for τ_0, one gets

$$\langle \hat{\tau} \rangle = \frac{12D\tau_{\text{p},1}}{5} \left(\frac{p_{\text{max}}}{p_{\text{trap}}}\right)^D. \tag{A.75}$$

---

[16] Equation (3.35) was calculated for the deterministic model, while Raman cooling corresponds to the exponential model. Taking into account the exponential model would simply introduce a prefactor of order one in eq. (A.74).

Appendix B

The Doppler case

B.1 Motivations

The case where the momentum of the atom is not confined by other external means is clearly very unfavourable for the cooling efficiency. For example, when the momentum performs a one-dimensional random walk outside the trap, the return time distribution decays for large return times $\hat{\tau}$ as $\hat{\tau}^{-3/2}$, which is precisely the same decay as the trapping time distribution itself when $\alpha = 2$. This in turn leads to the fact that the fraction of cooled atoms $f(\theta)$ tends to a certain constant (less than one) for large interaction times θ (see Section 5.2.5). The Doppler effect makes the situation much worse. Because of the frequency shift $\Delta\omega = kp/M$ the rate at which the atom can change its momentum decays as p^{-2} for large momentum (this comes from the Lorentzian line shape of the resonance), slowing down the motion for large p (see eq. (A.18)). Hence, the return times can only be longer because of this effect. Actually, as we are now going to show, the return time distribution (for a one-dimensional geometry) has a slower asymptotic decay for large $\hat{\tau}$, since it decays as $\hat{\tau}^{-5/4}$. Hence, from a practical point of view, this situation is not very interesting. However, from a theoretical point of view, it is important since it corresponds to a case where the fraction of trapped atoms decays to zero for large θ in a non-trivial way, which serves as an interesting testing ground both for the theoretical and numerical methods used in the present context. Here, we show how the tail of the return time distribution can be obtained exactly, using a remarkable mapping of the problem onto a quantum harmonic oscillator.

B.2 Setting the stage

The time $\hat{\tau}$ spent outside the trapping region after n steps can be written as:

$$\hat{\tau}(n) = \sum_{j=1}^{n} u_j \tau[p(j)]_{p(0)=p(n+1)\simeq 0}, \tag{B.1}$$

172

where n is the total number of absorbed and emitted photons, each of which changes the momentum by a random amount, before the atom comes back to the 'trap' $p \simeq 0$. We consider here the realistic exponential time model (see Section 3.3.1, p. 30). Therefore, the time between each jump can be written as the product of an exponential random variable u_j of mean equal to one, and a factor $\tau[p] = 1/R(p)$, which is the average time needed for the atom to change its momentum knowing that its current momentum is p. The latter quantity is given, in the Doppler case, by eq. (A.30):

$$\tau[p] = \tau_0 \left(\frac{p}{p_D}\right)^2 \qquad (p \text{ large}). \tag{B.2}$$

It is important to realize that the dynamics of the momentum is still a one-dimensional random walk as a function of the number of steps n (rather than the real time t). The Central Limit Theorem allows one to write the probability $\mathcal{P}(\mathcal{C}_n)$ to observe a certain path $\mathcal{C}_n = \{p(0), p(1), \ldots, p(n) = p_f\}$ with a very large number of steps n in the following Gaussian form:

$$\mathcal{P}(\mathcal{C}_n) = \frac{1}{Z_n} \exp\left\{-\frac{1}{2\Delta p^2} \sum_{j=1}^{n} [p(j) - p(j-1)]^2\right\}, \tag{B.3}$$

where Δp^2 is the variance of the momentum jump and Z_n the normalization factor. (Higher cumulants are negligible in the large time, large n, limit that we shall consider below.)

Let us now introduce an auxiliary function $G_n(s, p_f)$ defined as:

$$G_n(s, p_f) = \overline{e^{-s\hat{\tau}(n)}}, \tag{B.4}$$

where the overline means the average over all paths such that $p(0) = 0$ and $p(n) = p_f$. The quantity $G_n(s, p_f = 0)$, corresponding to paths ending in the trap, can thus be interpreted as the Laplace transform of the probability distribution of $\hat{\tau}(n)$, considered as first return times, for a given n. More explicitly, $G_n(s, p_f)$ reads:

$$G_n(s, p_f) = \int_{p(0)=0}^{p(n)=p_f} \mathcal{D}p(j) \int \prod_{j=1}^{n} du_j \, e^{-u_j}$$

$$\times \exp -\left\{\frac{1}{2\Delta p^2} \sum_{j=1}^{n} (p(j) - p(j-1))^2 + s \sum_{j=1}^{n} u_j \tau[p(j)]\right\} \tag{B.5}$$

where $\mathcal{D}p$ is the functional measure (that includes the normalization factor Z_n).

The integral over the u_j can be performed, leading to:

$$G_n(s, p_f) = \int_{p(0)=0}^{p(n)=p_f} \mathcal{D}p(j)$$

$$\times \exp - \left\{ \frac{1}{2\Delta p^2} \sum_{j=1}^{n} (p(j) - p(j-1))^2 + \sum_{j=1}^{n} \log(1 + s\tau[p(j)]) \right\}.$$

$$(B.6)$$

Since we are interested in the tail of the distribution of $\hat{\tau}$, we will consider the small-s limit in the previous expression, for which one can expand the logarithm as:

$$\log(1 + s\tau[p(j)]) \simeq s\tau[p(j)]. \tag{B.7}$$

This means that we could have considered the deterministic hopping time model (where $u_j = 1$), or any model where the average of u is equal to one, with exactly the same results.

B.3 Feynman path integral and mapping to the harmonic oscillator

One can also justify that large $\hat{\tau}$ values correspond to a large number of steps $n \gg 1$ outside the trap, so that the above discrete sums can safely be replaced by integrals where n varies continuously. Furthermore, $n \gg 1$ means that the random walks reach large momenta, such that eq. (B.2) is asymptotically valid. Hence one has:

$$G_n(s, p_f) = \int_{p(0)=0}^{p(n)=p_f} \mathcal{D}p \exp - \left\{ \frac{1}{2\Delta p^2} \int_0^n dn' \left(\frac{\partial p}{\partial n'} \right)^2 + s\tau_0 p_D^{-2} \int_0^n dn' \, p(n')^2 \right\}.$$

$$(B.8)$$

If one identifies $n \to it$, $n' \to it'$ and $p \to x$, the right-hand side of the above expression takes the familiar form of a Feynman path integral (with $\hbar = 1$):

$$\int_{x(0)=0}^{x(t)=x_f} \mathcal{D}x \exp i\mathcal{S}(\{x\}), \tag{B.9}$$

where the action is that of a harmonic oscillator:

$$\mathcal{S}(\{x\}) = \int_0^t dt' \left[\frac{m}{2} \left(\frac{dx}{dt'} \right)^2 - \frac{m\omega^2}{2} x^2 \right], \tag{B.10}$$

with the following identification:

$$\Delta p^2 \to \frac{1}{m}, \qquad s\tau_0 p_D^{-2} \to \frac{m\omega^2}{2}. \tag{B.11}$$

Therefore, using the standard results on Feynman path integrals (see, e.g. [FeH65]), one can identify G in eq. (B.8) with the propagator $\langle x_f | e^{-iHt} | 0 \rangle$ of a quantum harmonic oscillator, with a Hamiltonian:

$$H = -\frac{1}{2m} \frac{\partial^2}{\partial x^2} + \frac{m}{2} \omega^2 x^2,$$

(B.12)

i.e. the probability amplitude that the oscillator starts from position $x_i = 0$ and reaches x_f after time t. We can thus write:

$$G_n(s, p_f) = \langle p_f | e^{-nH} | 0 \rangle.$$

(B.13)

B.4 Back to the return time probability

Let us now introduce the probability $P_{\text{trap}}(\hat{\tau})$ of being at time $\hat{\tau}$ within a region of size $2p_{\text{trap}}$ around $p = 0$. Using the fact that paths with $|p_f| \leq p_{\text{trap}}$ after an arbitrary number of steps n contribute to $P_{\text{trap}}(\hat{\tau})$, one can write:

$$P_{\text{trap}}(\hat{\tau}) = 2p_{\text{trap}} \sum_{n=1}^{\infty} \overline{\delta(\hat{\tau}(n) - \hat{\tau})}.$$

(B.14)

Therefore, the Laplace transform of $P_{\text{trap}}(\hat{\tau})$ reads:

$$\mathcal{L}P_{\text{trap}}(s) = 2p_{\text{trap}} \sum_{n=1}^{\infty} \int d\hat{\tau} \, e^{-s\hat{\tau}} \overline{\delta(\hat{\tau}(n) - \hat{\tau})} = 2p_{\text{trap}} \sum_{n=1}^{\infty} \overline{e^{-s\hat{\tau}(n)}}.$$

(B.15)

Using the very definition of G_n, eq. (B.4), one therefore obtains the following identity:

$$\mathcal{L}P_{\text{trap}}(s) = 2p_{\text{trap}} \sum_{n=1}^{\infty} G_n(s, p_f = 0).$$

(B.16)

Finally, replacing the sum over n by an integral and using eq. (B.13) with $p_f = 0$, we obtain:

$$\mathcal{L}P_{\text{trap}}(s) \simeq 2p_{\text{trap}} \langle 0 | H^{-1} | 0 \rangle = 2p_{\text{trap}} \sum_{k=0}^{\infty} \frac{|\psi_k(0)|^2}{E_k},$$

(B.17)

where

$$E_k = \omega \left(k + \tfrac{1}{2} \right), \qquad |\psi_k(0)|^2 = \sqrt{\frac{m\omega}{\pi}} \frac{k!}{2^k ((k/2)!)^2}$$

(B.18)

are respectively the energy of the kth state of the harmonic oscillator and the squared modulus of the corresponding value of the wave function at the origin (see [LaL77]), valid for even k (while $\psi_k(0) = 0$ for odd k). Using the above 'dictionary' (B.11) between the harmonic oscillator and the parameters of the present problem, we find:

$$\mathcal{L}P_{\text{trap}}(s) = 2p_{\text{trap}} \left(\frac{s\Delta p^6 \tau_0}{8 p_D^2} \right)^{-1/4} \sum_{q=0}^{\infty} \frac{1}{2\sqrt{\pi}} \frac{(2q)!}{2^{2q} q!^2 (2q + \frac{1}{2})}, \qquad (B.19)$$

where we have set $k = 2q$. Now, the probability of the *first return* in the trap $\hat{P}(\hat{\tau})$ is easily expressed in terms of $P_{\text{trap}}(t)$ with Laplace transforms, as was done in Chapter 3:

$$\mathcal{L}P_{\text{trap}}(s) = \frac{1}{1 - \mathcal{L}[\hat{P}](s)}, \qquad (B.20)$$

which gives:

$$\mathcal{L}\hat{P}(s) = 1 - 1.3585 \ldots \frac{1}{2p_{\text{trap}}} \left(\frac{s\Delta p^6 \tau_0}{8 p_D^2} \right)^{1/4}. \qquad (B.21)$$

Using the results of Chapter 4, we can recognize the Laplace transform of the distribution of first return times which decays for large $\hat{\tau}$ as:

$$\hat{P}(\hat{\tau}) \underset{\hat{\tau}\to\infty}{\simeq} 0.3296 \ldots \frac{\Delta p^{3/2}}{p_{\text{trap}} p_D^{1/2}} \frac{\tau_0^{1/4}}{4\hat{\tau}^{5/4}}, \qquad (B.22)$$

corresponding to a broad distribution with an index which we called $\hat{\mu} = \frac{1}{4}$ in the main text (this in turn leads to an asymptotic decay of the proportion of trapped atoms as $\theta^{-1/4}$). Note that the time $\hat{\tau}_b$ defined in eq. (3.49) is thus given by:

$$\hat{\tau}_b = (0.3296 \ldots)^4 \frac{\Delta p^6}{p_{\text{trap}}^4 p_D^2} \tau_0. \qquad (B.23)$$

We have checked the above results, including the value of the prefactors, by performing a direct numerical simulation of the problem. Our numerical results are in good agreement with the above predictions: with a histogram of only 3000 points, the best fit to a $\hat{\tau}^{-5/4}$ tail for $\hat{P}(\hat{\tau})$ leads to a numerical constant of $\simeq 0.35$ for the tail amplitude, instead of the above cited value 0.3296... Note that the simulation was done with a uniform distribution of jump sizes instead of a Gaussian distribution for which the above calculation would be exact. However, as expected, the detailed shape of the jump size distribution is indeed irrelevant for large $\hat{\tau}$.

This completes our detailed analysis of the return time distribution in the presence of Doppler shift, which from a theoretical point of view is quite remarkable since it leads to a quantum oscillator harmonic problem which can be solved completely.

Appendix C

The special case $\mu = 1$

In the special case $\mu = 1$ where the average value of τ just diverges, the results given in the main text must be corrected by some logarithmic factor. Indeed, the Laplace transform of $P(\tau)$ can be expressed as:

$$\int_0^\infty d\tau \, P(\tau)(e^{-s\tau} - 1 + 1) = 1 + \int_0^\infty \frac{du}{s} P\left(\frac{u}{s}\right)(e^{-u} - 1). \tag{C.1}$$

However, direct substitution of the asymptotic form of $P(\tau)$ in the above formula leads to a logarithmic divergence for *small* u. The origin of this divergence is the fact that below a certain τ^* of the order of τ_b, it is certainly unjustified to replace $P(\tau)$ by τ_b/τ^2. The integral should thus rather be split into two parts, the first part reading, to leading order in s[1]:

$$\tau_b s \int_{\tau^* s}^\infty du \, u^{-2}(e^{-u} - 1) = \tau_b s \left[Ei(-\tau^* s) + \frac{\exp(-\tau^* s) - 1}{\tau^* s} \right]$$

$$\underset{s \to 0}{\simeq} \tau_b s (\log(\tau^* s)) + O(s) \tag{C.2}$$

where Ei is the Exponential–Integral function, defined as (see [GrR94] for a definition and for the small-argument expansion of this function):

$$Ei(x) = \int_x^\infty du \, u^{-1} e^{-u}. \tag{C.3}$$

The second part of the integral, corresponding to $\tau < \tau^*$, is given, at small s, by:

$$-s \int_0^{\tau^*} d\tau \, \tau P(\tau), \tag{C.4}$$

which therefore also contributes to order s. One thus finds:

$$\mathcal{L}P(s) = 1 - \tau_b s |\log(\tau^* s)| - A_1 \tau_b s + \cdots \tag{C.5}$$

---

[1] The subleading corrections to the τ^{-2} asymptotic behaviour of $P(\tau)$ also contribute (to order s) to the Laplace transform.

where A_1 is a certain numerical constant, which depends on the detailed form of $P(\tau)$.

Let us now show how this logarithmic correction affects the results for the sprinkling distribution $S(t)$. The relation between Laplace transforms:

$$\mathcal{L}S(s) = \frac{\mathcal{L}P(s)}{1 - \mathcal{L}P(s)} \tag{C.6}$$

of course still holds, which yields the following small-s behaviour of $\mathcal{L}S$:

$$\mathcal{L}S(s) \simeq \frac{1}{s\tau_b|\log(\tau^* s)|}. \tag{C.7}$$

By continuity with the case $\mu < 1$, $S(t)$ is expected to decay with t for $\mu = 1$, but presumably only logarithmically. Let us thus look for the Laplace transform of a function decaying for large t as $Z/[\log(t/\tau^*)]^\varsigma$. The Laplace transform of this function can be written as:

$$\mathcal{L}\frac{Z}{(\log(\frac{t}{\tau^*}))^\varsigma} = \frac{Z}{s} \int_0^\infty du \, \frac{e^{-u}}{|-\log(s\tau^*) + \log u|^\varsigma}. \tag{C.8}$$

For very small $s\tau^*$, one has $|\log u| \ll |\log(s\tau^*)|$ nearly everywhere in the integration domain. Therefore:

$$\mathcal{L}\frac{Z}{(\log(\frac{t}{\tau^*}))^\varsigma} = \frac{Z}{s|\log(s\tau^*)|^\varsigma} - \frac{\varsigma Z}{s|\log(s\tau^*)|^{\varsigma+1}} \int_0^\infty du \, e^{-u} \log u$$

$$+ O\left(\frac{1}{s|\log(s\tau^*)|^{\varsigma+2}}\right). \tag{C.9}$$

Comparing this with eq. (C.7), we thus find that $Z = 1/\tau_b$ and $\varsigma = 1$. This means that:

$$S(t) \underset{t\to\infty}{\simeq} \frac{1}{\tau_b \log(\frac{t}{\tau^*})}, \tag{C.10}$$

although one should bear in mind that the relative corrections to this result only decay as $1/\log t$ for large times.

The sprinkling distribution is thus of order $1/\tau_b$, as would be the case if $\mu > 1$, but with a (very weak) dependence on t which is reminiscent of the case $\mu < 1$. Strictly speaking, though, time translational invariance is already lost for $\mu = 1$.

When two types of traps compete, one of which is characterized by $\mu = 1$ and the other one by $\hat{\mu} > 1$, then the previous result on the sprinkling distribution remains unchanged, up to subdominant terms which are different. On the contrary, if $\hat{\mu} < 1$, the results reported in Chapter 6 are valid, although the subdominant terms, obviously, contain logarithms.

Let us now investigate the fraction of trapped atoms in the case where the trapping time distribution is such that $\mu = 1$, while the average time spent outside the trap is finite and equal to $\langle \hat{\tau} \rangle$. In this case, the small-s expansion of $\mathcal{L} f_{\text{trap}}$ reads:

$$\mathcal{L} f_{\text{trap}}(s) = \frac{1}{s} - \frac{\langle \hat{\tau} \rangle}{s \tau_b |\log(\tau^* s)|} + \cdots \qquad (C.11)$$

which shows that in this marginal case too, $f_{\text{trap}}(\theta) \to 1$ for $\theta \to \infty$. However, the convergence towards one is extremely slow since, using eq. (C.9) above:

$$f_{\text{trap}}(\theta) \underset{\theta \to \infty}{\sim} 1 - \frac{\langle \hat{\tau} \rangle}{\tau_b \log(\frac{\theta}{\tau^*})} + \cdots . \qquad (C.12)$$

Again, the behaviour of $f_{\text{trap}}(\theta)$ in the case $\mu = 1$ has features in common with both the cases $\mu < 1$ and $\mu > 1$: mathematically, $f_{\text{trap}}(\infty) = 1$; however, even for very large values of θ, $f_{\text{trap}}(\theta)$ looks very much as if it had saturated to a value < 1!

Finally, the *height* of the momentum distribution at $p = 0$, which we denote as $h(\theta)$ in the text, is simply obtained by integrating $S(t)$ up to time θ. For large θ, the integral is dominated by the neighbourhood of the upper bound θ, and thus $h(\theta) \propto \theta / \log(\theta/\tau^*)$. More precisely, the full momentum distribution can be written, for large θ/τ_0, in a scaling form similar to eqs. (6.31), (6.42) and (6.49):

$$\pi(p, \theta) = h(\theta)\, \mathcal{G}_1 \left(\frac{p}{p_\theta} \right) = h(\theta)\, \mathcal{G}_1 \left[\frac{p}{p_0} \left(\frac{\theta}{\tau_0} \right)^{1/\alpha} \right] \qquad (C.13)$$

with

$$h(\theta) = \frac{1}{C_D\, p_{\text{trap}}^D} \frac{\theta}{\tau_b \log(\frac{\theta}{\tau^*})}, \qquad (C.14)$$

$q = p/p_\theta$ and

$$\mathcal{G}_1(q) = q^{-\alpha} \left[1 - \exp(-q^\alpha) \right] \qquad (C.15)$$

for the exponential model. The form of the scaling function is therefore identical to the case $\mu > 1$ (see eq. (6.45)).

References

[AAK88] A. Aspect, E. Arimondo, R. Kaiser, N. Vansteenkiste, and C. Cohen-Tannoudji, Laser cooling below the one-photon recoil energy by velocity-selective coherent population trapping, *Phys. Rev. Lett.* **61**, p. 826–829 (1988).

[AAK89] A. Aspect, E. Arimondo, R. Kaiser, N. Vansteenkiste, and C. Cohen-Tannoudji, Laser cooling below the one-photon recoil energy by velocity-selective coherent population trapping: theoretical analysis, *J. Opt. Soc. Am. B* **6**, p. 2112–2124 (1989).

[AbS70] M. Abramowitz and I.A. Stegun, *Handbook of Mathematical Functions*, Ninth Edition, Dover, New York (1970).

[AdR97] C.S. Adams and E. Riis, Laser cooling and trapping of neutral atoms, *Progr. Quant. Electron.* **21**, p. 1–79 (1997).

[AGM76] G. Alzetta, A. Gozzini, L. Moi, and G. Orriols, An experimental method for the observation of R.F. transitions and laser beat resonances in oriented Na vapour, *Nuovo Cimento* **36B**, p. 5–20 (1976).

[AlK92] V.A. Alekseev and D.D. Krylova, Nearly absolute one-dimensional cooling of helium atoms by two oppositely directed monochromatic plane waves, *JETP Lett.* **55**, p. 321–325 (1992) (*Pis'ma Zh. Eksp. Teor. Fiz.* **55**, p. 321–324 (1992)).

[AlK93] V.A. Alekseev and D.D. Krylova, Coherent population trapping: asymptotical behaviour of the atomic density matrix, *Opt. Commun.* **95**, p. 319–326 (1993).

[AlK96] V.A. Alekseev and D.D. Krylova, Fraction number of trapped atoms and velocity distribution function in sub-recoil laser cooling scheme, *Opt. Commun.* **124**, p. 568–578 (1996).

[APS92] E. Arimondo, W.D. Phillips, and F. Strumia (editors), *Laser Manipulation of Atoms and Ions*, Proceedings of the 1991 Varenna Summer School, North-Holland, Amsterdam (1992).

[Ari91] E. Arimondo, *Velocity-Selective Coherent Population Trapping in One and Two Dimensions*, Proceedings of the 1991 Varenna Summer School, edited by E. Arimondo, W.D. Phillips and F. Strumia, p. 191–224, North-Holland, Amsterdam (1992).

[Ari96] E. Arimondo, Relaxation processes in coherent-population trapping, *Phys. Rev. A* **54**, p. 2216–2223 (1996).

[ArO76] E. Arimondo and G. Orriols, Nonabsorbing atomic coherences by coherent two-photon transitions in a three-level optical pumping, *Lett. Nuovo Cimento* **17**, p. 333–338 (1976).

[BaB00]   O.E. Barndorff-Nielsen and F.E. Benth, *Laser cooling and stochastics*, to appear in *State of the Art in Probability and Statistics; Festschrift for Willem R. van Zwet*, M.C.M. de Gunst, C.A.J. Klaassen, A.W. van der Vaart, (eds.), Lecture Notes–Monograph Series (Inst. Math. Statist., Beachwood, OH).

[Bak96]   P. Bak, *How Nature Works*, Copernicus, Springer-Verlag, New-York (1996).

[Bar95]   F. Bardou, *Refroidissement laser sub-recul par résonances noires*, PhD Thesis, University of Paris XI Orsay (1995).

[BBE94]   F. Bardou, J.P. Bouchaud, O. Emile, A. Aspect, and C. Cohen-Tannoudji, Subrecoil laser cooling and Lévy flights, *Phys. Rev. Lett.* **72**, p. 203–206 (1994).

[BBJ00]   O.E. Barndorff-Nielsen, F.E. Benth, and J.L. Jensen, Markov jump processes with a singularity, *Adv. Appl. Prob.* **32**, p. 779–799 (2000).

[BCK97]   J.P. Bouchaud, L.F. Cugliandolo, J. Kurchan, and M. Mézard, *Out of equilibrium dynamics in spin-glasses and other glassy systems*, in *Spin-Glasses and Random Fields*, edited by A. P. Young, Series on Directions in Condensed Matter Physics **12**, World Scientific, Singapore (1997).

[BEC]   For recent general references on Bose–Einstein condensation, see: S. Martellucci, A.C. Chester, A. Aspect and M. Inguscio, *Bose–Einstein Condensates and Atom Lasers*, Proceedings of the 27th Course of the International School of Quantum Electronics (Erice, 19–24 October 1999); A. Aspect J. Dalibard, *Bose–Einstein Condensate and Atom Lasers*, Special issue of the Comptes Rendus de l'Académie des Sciences, C.R. Acad. Sci. 20, série IV (2000), and M. Inguscio, S. Stringari, and C.E. Wieman, *Bose–Einstein Condensation in Atomic Gases*, Proceedings of the International School Enrico Fermi, Course CXL, IOS Press (1999).

[BoC97]   M. Boguna and A. Corral, Long-tailed trapping times and Lévy flights in a self-organized critical granular system, *Phys. Rev. Lett.* **78**, p. 4950–4953 (1997).

[BoD95]   J.P. Bouchaud and D.S. Dean, Aging in Parisi's Tree, *J. Physique I (France)* **5**, p. 265–286 (1995).

[BoG90]   J.P. Bouchaud and A. Georges, Anomalous diffusion in disordered media: statistical mechanisms, models and physical applications, *Phys. Rep.* **195**, p. 127–293 (1990).

[BoP00]   J.P. Bouchaud and M. Potters, *Theory of Financial Risks*, Cambridge University Press, Cambridge, 2000.

[Bou92]   J.P. Bouchaud, Weak ergodicity breaking and aging in disordered systems, *J. Physique I (France)* **2**, p. 1705–1713 (1992)

[Bou00]   J.P. Bouchaud, in *Soft and Fragile Matter*, eds M.E. Cates and M.R. Evans, Institute of Physics Publishing (Bristol and Philadelphia), 2000.

[BrP96]   H.P. Breuer and F. Petruccione, Hilbert space path integral representation for the reduced dynamics of matter in thermal radiation fields, *J. Phys. A: Math. Gen.* **29**, p. 7837–7853 (1996).

[BSL94]   F. Bardou, B. Saubaméa, J. Lawall, K. Shimizu, O. Emile, C. Westbrook, A. Aspect, and C. Cohen-Tannoudji, Sub-recoil laser cooling with precooled atoms, *C. R. Acad. Sci. Paris* **318**, Série II, p. 877–885 (1994).

[BWS01]   T. Binnewies, G. Wilpers, U. Sterr, F. Riehle, J. Helmcke, T.E. Mehlstäubler, E.M. Rasel, and W. Ertmer, *Doppler cooling and trapping on forbidden transitions*, preprint http://www.lanl.gov/abs/physics/0105069 (2001).

[CaD91]   Y. Castin and J. Dalibard, Quantization of atomic motion in optical molasses, *Europhys. Lett.* **14**, p. 761–766 (1991).

[CaM95]   Y. Castin and K. Mølmer, Monte-Carlo wave-function analysis of 3D optical molasses, *Phys. Rev. Lett.* **74**, p. 3772–3775 (1995).

[Car93]   H.J. Carmichael, *An Open Systems Approach to Quantum Optics*, Lecture Notes in Physics m18, Springer-Verlag, New York (1993).

[CBA91]   C. Cohen-Tannoudji, F. Bardou, and A. Aspect, *Review of fundamental processes in laser cooling*, Proceedings of Laser Spectroscopy X (Font-Romeu, 1991), edited by M. Ducloy, E. Giacobino, and G. Camy, p. 3–14, World Scientific, Singapore (1992).

[CDG88]   C. Cohen-Tannoudji, J. Dupont-Roc, and G. Grynberg, *Processus d'interaction entre photons et atomes*, InterEditions et Editions du CNRS (1988). English version: *Atom–Photon Interactions. Basic Processes and Applications*, Wiley-Interscience, Singapore (1992).

[CHB85]   S. Chu, L. Hollberg, J.E. Bjorkholm, A. Cable, and A. Ashkin, Three-dimensional viscous confinement and cooling of atoms by resonance radiation pressure, *Phys. Rev. Lett.* **55**, p. 48–51 (1985).

[Chu98]   S. Chu, The manipulation of neutral particles, *Rev. Mod. Phys.* **70**, p. 685–706 (1998).

[CoD86]   C. Cohen-Tannoudji and J. Dalibard, Single-atom laser spectroscopy. Looking for dark periods in fluorescence light, *Europhys. Lett.* **1**, p. 441–448 (1986).

[Coh90]   C. Cohen-Tannoudji, *Atomic motion in laser light*, in *Fundamental Systems in Quantum Optics*, J. Dalibard, J.M. Raimond, and J. Zinn-Justin eds, Les Houches session LIII (1990), p. 1–183, North-Holland, Amsterdam (1992).

[Coh92]   C. Cohen-Tannoudji, Cours au Collège de France 1991–1992, lecture notes, available www.ens.fr/cct.

[Coh96]   C. Cohen-Tannoudji, Cours au Collège de France 1995–1996, lecture notes available www.ens.fr/cct.

[Coh98]   C. Cohen-Tannoudji, Manipulating atoms with photons, *Rev. Mod. Phys.* **70**, p. 707–719 (1998).

[CoK85]   R.J. Cook and H.J. Kimble, Possibility of direct observation of quantum jumps, *Phys. Rev. Lett.* **54**, p. 1023–1026 (1985).

[CoP90]   C. Cohen-Tannoudji, W. D. Phillips, New mechanisms for laser cooling, *Phys. Today* **43**, p. 33–40 (1990).

[CoR00]   G. Combe and J.N. Roux, Strain versus stress in granular materials: A devil's staircase, *Phys. Rev. Lett.* **85**, 3628–3631 (2000).

[CZA93]   C. Cohen-Tannoudji, B. Zambon, and E. Arimondo, Quantum-jump approach to dissipative processes: application to amplification without inversion, *J. Opt. Soc. Am. B* **10**, p. 2107–2120 (1993).

[DaC89]   J. Dalibard and C. Cohen-Tannoudji, Laser cooling below the Doppler limit by polarization gradients: simple theoretical models, *J. Opt. Soc. Am. B* **6**, p. 2023–2045 (1989).

[DCM92]   J. Dalibard, Y. Castin, and K. Mølmer, Wave-function approach to dissipative processes in quantum optics, *Phys. Rev. Lett.* **68**, p. 580–583 (1992).

[Der94]   B. Derrida, *Non-self averaging effects in sum of random variables*, in *On Three Levels*, Edited by Fannes *et al.*, page 125, Plenum Press, New York (1994).

[DLK94]   Nir Davidson, Heun-Jin Lee, Mark Kasevich, and Steven Chu, Raman cooling of atoms in two and three dimensions, *Phys. Rev. Lett.* **72**, p. 3158–3161

(1994).

[DrD69]   G.W.F. Drake and A. Dalgarno, Intercombination oscillator strengths in the
           helium sequence, *Astrophys. J.* **157**, p. 459–462 (1969).

[Dyn61]   E.B. Dynkin, Some limit theorems for sums of independent random variables
           with infinite mathematical expectations, *Selected Trans. Math. Statist. Prob.*
           1, p. 171–189 (1961) IMS-AMS.

[DZR92]   R. Dum, P. Zoller, and H. Ritsch, Monte Carlo simulation of the atomic master
           equation for spontaneous emission, *Phys. Rev. A* **45**, p. 4879–4887 (1992).

[ESW96]   T. Esslinger, F. Sander, M. Weidemüller, A. Hemmerich, and T. W. Hänsch,
           Subrecoil laser cooling with adiabatic transfer, *Phys. Rev. Lett.* **76**,
           p. 2432–2435 (1996).

[FeH65]   R.P. Feynman and A.R. Hibbs, *Quantum Mechanics and Path Integrals*,
           McGraw Hill, New York (1965).

[Fel71]   W. Feller, *An Introduction to Probability Theory and Its Applications,
           Volume II*, John Wiley & Sons, New York (1971).

[FZA95]   A. Fioretti, B. Zambon, and E. Arimondo, Limits of velocity-selective
           coherent population trapping, *Quantum Semiclass. Opt.* **7**, p. 751–756
           (1995).

[GnK54]   B.V. Gnedenko and A.N. Kolmogorov, *Limit Distributions for Sums of
           Independent Random Variables*, Addison-Wesley, London (1954).

[GoL01]   C. Godrèche and J.M. Luck, *Statistics of the occupation time of renewal
           processes*, e-print cond-mat/0010428 and *J. Stat. Phys.* **104**, to appear
           (2001).

[GPS95]   E. Goldstein, P. Pax, K.J. Schernthanner, B. Taylor, and P. Meystre, Influence
           of the dipole–dipole interaction on velocity-selective coherent population
           trapping, *Appl. Phys. B* **60**, p. 161–167 (1995).

[GrR94]   I.S. Gradshteyn and I.M. Ryzhik, *Table of Integrals, Series and Products*,
           Fifth Edition, A. Jeffrey (ed.), Academic Press, New York (1994).

[Gum58]   E.J. Gumbel, *Statistics of Extremes*, Columbia University Press, New York
           (1958).

[HaS75]   T.W. Hänsch and A.L. Schawlow, Cooling of gases by laser radiation, *Opt.
           Commun.* **13**, p. 68–69 (1975).

[Hil75]   B.M. Hill, A simple general approach to inference about the tail of a
           distribution, *Ann. Stat.* **3**, p. 1163–1174 (1975).

[HLO00]   J. Hack, L. Liu, M. Olshanii, and H. Metcalf, Velocity-selective coherent
           population trapping of two level atoms, *Phys. Rev. A* **62**, p.
           013405.1–013405.4 (2000).

[IIK00]   T. Ido, Y. Isoya, and H. Katori, Optical-dipole trapping of Sr atoms at a high
           phase-space density, *Phys. Rev. A* **61**, p. 061403.1–061403.4 (2000).

[KaC92]   M. Kasevich and S. Chu, Laser cooling below a photon recoil with three-level
           atoms, *Phys. Rev. Lett.* **69**, p. 1741–1744 (1992).

[KII99]   H. Katori, T. Ido, Y. Isoya, and M. Kuwata-Gonokami, Magneto-optical
           trapping and cooling of strontium atoms down to the photon recoil
           temperature, *Phys. Rev. Lett.* **82**, p. 1116–1119 (1999).

[KSP97]   S. Kulin, B. Saubaméa, E. Peik, J. Lawall, T. W. Hijmans, M. Leduc, and C.
           Cohen-Tannoudji, Coherent manipulation of atomic wave packets by
           adiabatic transfer, *Phys. Rev. Lett.* **78**, p. 4185–4188 (1997).

[KSW97]   H. Katori, S. Schlipf, and H. Walther, Anomalous dynamics of a single ion in
           an optical lattice, *Phys. Rev. Lett.* **79**, p. 2221–2224 (1997).

[KSZ96]   J. Klafter, M.F. Shlesinger, and G. Zumofen, Beyond Brownian motion, *Phys.*

Today **49**, p. 33–39 (February 1996).

[LAK96]   H.J. Lee, C. Adams, M. Kasevich, and S. Chu, Raman cooling of atoms in an optical dipole trap, *Phys. Rev. Lett.* **76**, p. 2658–2661 (1996).

[LaL77]   L.D. Landau and E.M. Lifshitz, *Quantum Mechanics*, Pergamon Press, New York (1977).

[Lam58]   J. Lamperti, An occupation time theorem for a class of stochastic processes, *Trans. Am. Math. Soc.* **88**, 380 (1958).

[LBS94]   J. Lawall, F. Bardou, B. Saubamea, K. Shimizu, M. Leduc, A. Aspect, and C. Cohen-Tannoudji, Two-dimensional subrecoil laser cooling, *Phys. Rev. Lett.* **73**, p. 1915–1918 (1994).

[Lev37]   P. Lévy, *Théorie de l'addition des variables aléatoires*, Gauthier-Villars, Paris (1937, also 1954).

[LJD77]   C.D. Lin, W.R. Johnson, and A. Dalgarno, Radiative decays of the $n = 2$ states of He-like ions, *Phys. Rev. A* **15**, p. 154–161 (1977).

[LKS95]   J. Lawall, S. Kulin, B. Saubamea, N. Bigelow, M. Leduc, and C. Cohen-Tannoudji, Three-dimensional laser cooling of helium beyond the single-photon recoil limit, *Phys. Rev. Lett.* **75**, p. 4194–4197 (1995).

[LPR89]   P.D. Lett, W.D. Phillips, S.L. Rolston, C.E. Tanner, R.N. Watts, and C.I. Westbrook, Optical molasses, *J. Opt. Soc. Am. B* **6**, p. 2084–2107 (1989).

[MaA91]   F. Mauri and E. Arimondo, Coherent trapping subrecoil cooling in two dimensions aided by a force, *Europhys. Lett.* **16**, p. 717–722 (1991).

[MaA92]   F. Mauri and E. Arimondo, Two dimension selective coherent population trapping controlled by a phase shift, *Appl. Phys. B* **54**, p. 420–427 (1992).

[MaM93]   B.G. Matisov and I.E. Mazets, Limit of laser cooling of atoms by velocity selective coherent population trapping, *J. Phys. B: At. Mol. Opt. Phys.* **26**, p. 3795–3802 (1993).

[Man82]   B.B. Mandelbrot, *The Fractal Geometry of Nature*, Freeman, San Francisco (1982).

[Man96]   B. Mandelbrot, Du hasard bénin au hasard sauvage, *Pour la Science*, p. 12–17 (April 1996).

[Man97]   B.B. Mandelbrot, *Fractals and Scaling in Finance. Discontinuity, Concentration, Risk*, Springer-Verlag, Berlin (1997).

[MaS99]   R. Mantegna and H.E. Stanley, *Introduction to Econophysics*, Cambridge University Press, Cambridge (1999).

[MCD93]   K. Mølmer, Y. Castin, and J. Dalibard, Monte Carlo wave-function method in quantum optics, *J. Opt. Soc. Am. B* **10**, p. 524–538 (1993).

[MDT94]   P. Marte, R. Dum, R. Taïeb, P. Zoller, M.S. Shahriar, and M. Prentiss, Polarization-gradient-assisted subrecoil cooling: Quantum calculations in one dimension, *Phys. Rev. A* **49**, p. 4826–4836 (1994).

[MeV99]   H. Metcalf and P. van der Straten, *Laser Cooling and Trapping*, Series: Graduate Texts in Contemporary Physics, Springer-Verlag, Berlin (1999).

[MEZ96]   S. Marksteiner, K. Ellinger, and P. Zoller, Anomalous diffusion and Lévy walks in optical lattices, *Phys. Rev. A* **53**, p. 3409–3430 (1996).

[MiR85]   V.G. Minogin, Yu.V. Rozhdestvenskiĭ, Coherent dragging of atomic populations in problems of resonant light pressure, *Sov. Phys. JETP* **61**, p. 1156–1159 (1985) (*Zh. Éksp. Teor. Fiz.* **88**, p. 1950–1957 (1985)).

[MKG94]   B.G. Matisov, E.A. Korsunsky, V. Gordienko, and L. Winholz, Dynamics and limits of laser cooling by velocity selective coherent population trapping, *Laser Phys.* **4**, p. 835–847 (1994).

[MoC96]   K. Mølmer and Y. Castin, Monte Carlo wavefunctions in quantum optics, *Quantum Semiclass. Opt.* **8**, p. 49–72 (1996).

[Mol91]   K. Mølmer, Limits of Doppler cooling in pulsed laser fields, *Phys. Rev. Lett.* **66**, p. 2301–2304 (1991).

[Mor95]   O. Morice, *Atomes refroidis par laser: du refroidissement subrecul à la recherche d'effets quantiques collectifs*, PhD Thesis, University of Paris VI (1995).

[MZL96]   G. Morigi, B. Zambon, N. Leinfellner, and E. Arimondo, Scaling laws in velocity-selective coherent-population-trapping laser cooling, *Phys. Rev. A* **53**, p. 2616–2626 (1996).

[NWS96]   M. Narascheuski, H. Wallis, and A. Schenzle, Quantum-statistical enhancement of spontaneous emission in velocity-selective coherent population trapping, *Phys. Rev. A* **54**, p. 677–690 (1996).

[OlM90]   M.A. Ol'shanii and V.G. Minogin, *Three-dimensional velocity-selective coherent atomic population trapping in resonant light fields*, Proceedings of the international workshop on 'Light induced kinetic effects on atoms, ions and molecules', eds. L. Moi, S. Gozzini, C. Gabanini, E. Arimondo and F. Strumia (Marciana Marina, Italy, 1990), p. 99–110, Ets Editrice, Pisa (1991).

[PaB99]   W. Paul and J. Baschnagel, *Stochastic Processes: From Physics to Finance*, Springer-Verlag, Berlin (1999).

[PHB87]   D. Pritchard, K. Helmerson, V.S. Bagnato, G.P. Lafyatis, and A.G. Martin, in Proceedings of the VIII Conference on Laser Spectroscopy, eds. S. Svandberg and W. Persson, p. 68, Springer-Verlag, Berlin (1987).

[Phi98]   W.D. Phillips, Laser cooling and trapping of neutral atoms, *Rev. Mod. Phys.* **70**, p. 721–741 (1998).

[PMA92]   F. Papoff, F. Mauri, and E. Arimondo, Transient velocity-selective coherent population trapping in one dimension, *J. Opt. Soc. Am. B* **9**, p. 321–331 (1992).

[PlK98]   M.B. Plenio and P.L. Knight, The quantum-jump approach to dissipative dynamics in quantum optics, *Rev. Mod. Phys.* **70**, p. 101–144 (1998).

[RBB95]   J. Reichel, F. Bardou, M. Ben Dahan, E. Peik, S. Rand, C. Salomon, and C. Cohen-Tannoudji, Raman cooling of cesium below 3 nK: new approach inspired by Lévy flight statistics, *Phys. Rev. Lett.* **75**, p. 4575–4578 (1995).

[RDC88]   S. Reynaud, J. Dalibard, and C. Cohen-Tannoudji, Photon statistics and quantum jumps; the picture of the dressed atom radiative cascade, *IEEE J. Quant. Electron.* **24**, p. 1395–1402 (1988).

[Rei96]   J. Reichel, *Refroidissement Raman et vols de Lévy: atomes de césium au nanokelvin*, PhD Thesis, University of Paris VI (1996).

[Rey83]   S. Reynaud, La fluorescence de résonance: Etude par la méthode de l'atome habillé; Resonance fluorescence: the dressed atom approach, *Ann. Physique France* **8**, p. 315–370 (1983).

[RSC01]   J. Reichel, C. Salomon, and C. Cohen-Tannoudji, *Optimizing Raman Cooling*, to be published.

[Sau98]   B. Saubaméa, *Refroidissement laser subrecul au nanoKelvin. Mesure directe de la longueur de cohérence. Nouveaux tests des statistiques de Lévy*, Thèse de doctorat de l'université Paris VI (1998).

[Sch98]   S. Schaufler, *Scaling Theory of 1D-VSCPT Laser Cooling*, PhD Thesis, University of Ulm (1998).

[SHK97]   B. Saubaméa, T.W. Hijmans, S. Kulin, E. Rasel, E. Peik, M. Leduc, and

C. Cohen-Tannoudji, Direct measurement of the spatial correlation function of ultracold atoms, *Phys. Rev. Lett.* **79**, p. 3146–3149 (1997).

[Shl88]   M.F. Shlesinger, Fractal time in condensed matter, *Ann. Rev. Phys. Chem.* **39**, p. 269–290 (1988).

[SHP93]   M.S. Shahriar, P.R. Hemmer, M.G. Prentiss, P. Marte, J. Mervis, D.P. Katz, N.P. Bigelow, and T. Cai, Continuous polarization-gradient precooling-assisted velocity-selective coherent population trapping, *Phys. Rev. A* **48**, p. R4035–R4038 (1993).

[SLC99]   B. Saubaméa, M. Leduc, and C. Cohen-Tannoudji, Experimental investigation of non-ergodic effects in subrecoil laser cooling, *Phys. Rev. Lett.* **83**, p. 3796–3799 (1999).

[SSY97]   S. Schaufler, W.P. Schleich, and V.P. Yakovlev, Scaling and asymptotic laws in subrecoil laser cooling, *Europhys. Lett.* **39**, p. 383–388 (1997).

[SZF95]   M.F. Shlesinger, G.M. Zaslavsky, and U. Frisch (Eds), *Lévy flights and related topics in physics*, Lecture Notes in Physics **450**, Springer-Verlag, Berlin (1995).

[TaH72]   H.Y.S. Tang and W. Happer, Experimental determination of the $2^3P_1 \rightarrow 1^1S_0$ forbidden decay rate in helium, *Bull. Am. Phys. Soc.* **17**, p. 476 (1972).

[UWR89] P.J. Ungar, D.S. Weiss, E. Riis, and S. Chu, Optical molasses and multilevel atoms: theory, *J. Opt. Soc. Am. B* **6**, p. 2058–2071 (1989).

[WaE89] H. Wallis and W. Ertmer, Broadband laser cooling on narrow transitions, *J. Opt. Soc. Am. B* **6**, p. 2211–2219 (1989).

[Wal95]   H. Wallis, Quantum theory of atomic motion in laser light, *Phys. Rep.* **255**, p. 203–287 (1995).

[Wei94]   G. Weiss, *Random Walk Theory and Applications*, North-Holland, Amsterdam (1994).

[WEO94] M. Weidemüller, T. Esslinger, M.A. Ol'shanii, A. Hemmerich, and T.W. Hänsch, A novel scheme for efficient cooling below the photon recoil limit, *Europhys. Lett.* **27**, p. 109–114 (1994).

[WHF95] H. Wu, M.J. Holland, and C.J. Foot, Quantum jump calculations of velocity-selective coherent population trapping in one and two dimensions, *J. Phys. B: At. Mol. Opt. Phys.* **28**, p. 5025–5042 (1995).

[WiD75]   D. Wineland and H. Dehmelt, Proposed $10^{14}\Delta\nu < \nu$ laser fluorescence spectroscopy on Tl^+ mono-ion oscillator III, *Bull. Am. Phys. Soc.* **20**, p. 637 (1975).

[Zas99]   G.M. Zaslavsky, Chaotic dynamics and the origin of statistical laws, *Phys. Today* p. 39–45 (August 1999).

[ZMW87] P. Zoller, M. Marte, and D.F. Walls, Quantum jumps in atomic systems, *Phys. Rev. A* **35**, p. 198–207 (1987).

Index of main notation

Latin symbols

Notation	Definition	Page
$\mathcal{A}_\mu$	prefactor of $P(\tau)$ depending on the type of distribution for $P(\tau\|p)$	33
C_D	volume of the unit sphere in D dimensions: $C_1 = 2$, $C_2 = \pi$, $C_3 = 4\pi/3$	32
$\mathcal{D}(\theta)$	phase space density in $\mathbf{p} = 0$ at time θ	74
$f_{\text{peak}}(\theta)$	fraction of cooled atoms ($p < p_\theta$) at time θ	74
$f_{\text{trap}}(\theta)$	proportion of trapped atoms ($p < p_{\text{trap}}$) at time θ	60, 71
$f_{\text{trap}}(\theta)\big\|_{\text{opt}}$	proportion of trapped atoms evaluated at the time θ for which $h(\theta)$ is optimized	134
F	multiplicative factor allowing VSCPT quantum calculations to give the unconfined model	153
$\mathcal{F}_p$	family of the three states $\{\|g_-\rangle_p, \|g_+\rangle_p, \|e\rangle_p\}$ coupled by absorption and stimulated emission in one-dimensional σ_+/σ_- VSCPT	146
$\mathcal{G}(q)$	rescaled momentum distribution: $\pi(p,\theta) = h(\theta)\,\mathcal{G}(p/p_\theta)$. Case $\langle\tau\rangle$ infinite and $\langle\hat{\tau}\rangle$ finite	76
$\widetilde{\mathcal{G}}(q)$	same definition as $\mathcal{G}(q)$ but with $\langle\tau\rangle$ and $\langle\hat{\tau}\rangle$ finite	81
$\hat{\mathcal{G}}(q)$	same definition as $\mathcal{G}(q)$ but with $\langle\tau\rangle$ and $\langle\hat{\tau}\rangle$ infinite	85
$h(\theta)$	height of the peak of cooled atoms: $h(\theta) = \mathcal{P}(\mathbf{p} = 0, \theta)$	73

189

Notation	Definition	Page
$h(\theta)\vert_{\mathrm{opt}}$	optimized height of the peak of cooled atoms	134
$\hbar k$	photon momentum	2
$\hat{H}_p$	effective Hamiltonian describing the reduced evolution within $\mathcal{F}_p$	148
k	photon wave-vector modulus	2
k_{B}	Boltzmann constant	8
$\mathcal{L}f(s)$	Laplace transform of any function $f(t)$	45
$L_\mu(\xi)$	completely asymmetric Lévy distribution of index μ	45
M	atomic mass	8
N	number of terms in a sum $T_N = \sum_{i=1}^{N} \tau_i$; in particular,	44
	number of trapping events during the interaction time θ	4, 23
N_{samp}	number of samples (atoms) used in quantum jump simulations	108
$\mathbf{p}$	atomic momentum	2
p	modulus of the atomic momentum: $p = \Vert \mathbf{p} \Vert$	25, 71
$p_{1/2}$	half-width of function $\tau(p)$ when $R_0 \neq 0$: $R_0 = \frac{1}{\tau_0} \left(\frac{p_{1/2}}{p_0} \right)^\alpha$	95
p_{D}	Doppler width: $k p_{\mathrm{D}}/M = \Gamma/2$	26
$p_{\mathrm{m}}(\theta)$	median momentum of the trapped atoms at time θ	73
p_0	width of the dip of the jump rate $R(p)$ in $p \simeq 0$	23, 25
p_{max}	wall in momentum space (cf. effect of friction forces)	25
p_{R}	single photon recoil: $p_{\mathrm{R}} = \hbar k$	2
p_{trap}	size of the momentum trap	23
p_θ	characteristic momentum at time θ: $R(p_\theta) \cdot \theta = 1$	70
$p_{\theta,\mathrm{opt}}$	characteristic momentum p_θ evaluated at the time θ for which $h(\theta)$ is optimized	133
$P(x)$	probability distribution (probability density) of any variable x	29

Notation	Definition	Page
$P_i(p_x)$	jump rate induced by the ith pulse in Raman cooling	164–166
$P(\tau)$	probability distribution of trapping times τ, for an atom landing anywhere in the trap	5, 24
$P(\tau\|p)$	probability distribution of the sojourn times $\tau(p)$ at momentum p (deterministic or exponential model)	30, 71
$\hat{P}(\hat{\tau})$	probability distribution of recycling times $\hat{\tau}$	5, 24
$\mathcal{P}(\mathbf{p})$	probability distribution of the atomic momentum vector $\mathbf{p}$	69, 72
$\mathcal{P}(p, \theta)$	probability distribution of the momentum modulus p at time θ	71
Q	atomic parameter determined by the specific laser configuration and appearing in the expression of τ_b	126, 127
$\hat{Q}$	atomic parameter determined by the specific laser configuration and appearing in the expression of $\langle \hat{\tau} \rangle$	127
$R(\mathbf{p})$	jump rate (i.e. fluorescence rate, photon scattering rate): $R(p) = 1/\langle \tau(p) \rangle$	2, 20
R_0	jump rate in $p = 0$: $R_0 = R(p = 0)$	26, 93
s	conjugate of a time variable through a Laplace transform	39, 46
$S(t)$	'sprinkling distribution' (renewal density)	55, 56
S_D	surface of unit sphere in D dimensions: $S_D = DC_D$	32
$S_E(t)$	sprinkling distribution of exit times, i.e. rate of exit from the trap	62
$S_R(t)$	sprinkling distribution of return times, i.e. rate of entry in the trap	62
t_l	last trapping time before $t = \theta$	62
T	effective temperature: $k_B T = \delta p^2 / M$	8
T_N	sum of any N independent positive random variables τ_i (in Chapter 4): $T_N = \sum_{i=1}^{N} \tau_i$; in particular, total trapping time during an interaction of duration θ (in other chapters)	44 24

Notation	Definition	Page
$\hat{T}_N$	total recycling time during an interaction of duration θ: $\hat{T}_N = \sum_{i=1}^{N} \hat{\tau}_i$	24
T_R	recoil temperature: $T_R = \hbar^2 k^2 / (M k_B)$	9
$V_D(p)$	volume of a sphere of radius p in D dimensions: $V_D(p) = C_D p^D$	32
$w(\theta)$	half-width at $e^{-1/2}$ of the peak of cooled atoms (idem δp): $\pi(p = w(\theta), \theta) = e^{-1/2} \pi(p = 0, \theta)$	70, 73
$W(\tau)$	waiting time distribution (delay function), i.e. distribution of the time intervals τ between two *successive* spontaneous emissions	14, 58
$Y(x)$	Heaviside function ($x < 0$: $Y(x) = 0$; $x \geq 0$: $Y(x) = 1$)	49, 72

Greek symbols

Notation	Definition	Page
α	exponent of p $(< p_0)$ in $R(p)$: $R(p) = \frac{1}{\tau_0}\left(\frac{p}{p_0}\right)^\alpha$	25
$\gamma(\beta, x)$	incomplete Gamma function: $\gamma(\beta, x) = \int_0^x e^{-u}u^{\beta-1}du$	31
Γ	natural width of the excited state e. Γ^{-1} is the radiative lifetime of e.	148
$\Gamma_{g\pm}(p)$	decay rates of states $|g_\pm\rangle_p$	152
Γ_j	decay rates of the three eigenmodes of $\hat{H}_p$: $\Gamma_3(p) \simeq \Gamma$, $\Gamma_1(p) = \Gamma_C(p)$, $\Gamma_2(p) = \Gamma_{NC}(p)$	148–152
$\Gamma(x)$	gamma function: $\Gamma(x) = \int_0^\infty t^{x-1}e^{-t}\,dt$	31, 59
$\tilde{\delta}$	effective detuning between the laser frequency ω_L and the atomic frequency ω_A, including the recoil frequency Ω_R: $\tilde{\delta} = \omega_L - \omega_A + \Omega_R$	148
δp	half-width at $e^{-1/2}$ of the peak of cooled atoms (idem $w(\theta)$)	8
Δp	root mean square (rms) step length of the momentum random walk	22, 28
θ	duration of the interaction between the atoms and the laser beams	3, 12
λ	optimization parameter appearing in the expressions of τ_b and $\langle\hat{\tau}\rangle$: $\lambda = (\Omega_1/\Gamma)^2$ for VSCPT, $\lambda = 1/(\Gamma\tau_{p,1})$ for Raman cooling	126–127
λ_A	wavelength associated with the atomic transition with frequency ω_A: $\lambda_A = 2\pi c/\omega_A$	148
$\lambda_{opt}(\theta)$	value of λ optimizing the height $h(\theta)$ of the momentum distribution after an interaction time θ	131
μ	exponent of a power-law probability distribution $P(\tau) \underset{\tau\to\infty}{\simeq} \mu\tau_b^\mu/(\tau^{1+\mu})$; in particular, exponent of the probability distribution of trapping times: $\mu = D/\alpha$	43 ; 33
$\hat{\mu}$	exponent of the probability distribution of recycling times (when power-law distributed)	41
$\pi(p, \theta)$	reduced momentum distribution, more precisely one-dimensional section of the three-dimensional momentum distribution	72

Index